Willard Nelson Clute

The flora of the upper Susquehanna and its tributaries

Willard Nelson Clute

The flora of the upper Susquehanna and its tributaries

ISBN/EAN: 9783337268442

Printed in Europe, USA, Canada, Australia, Japan

Cover: Foto ©berggeist007 / pixelio.de

More available books at **www.hansebooks.com**

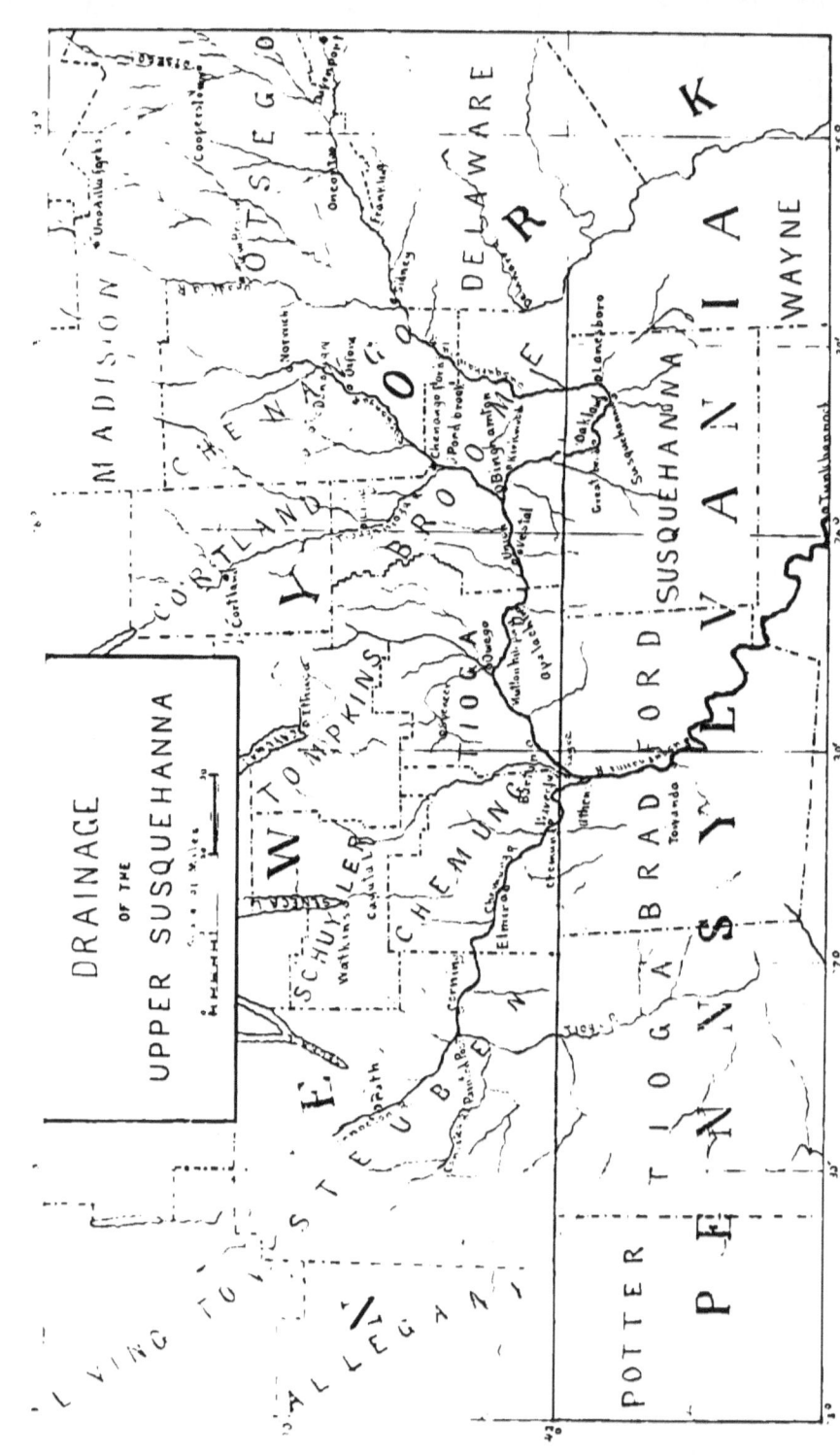

THE FLORA OF THE
UPPER SUSQUEHANNA
AND ITS TRIBUTARIES

By WILLARD NELSON CLUTE

Binghamton, N. Y.
Willard N. Clute & Co.
1898

PREFACE.

THE purpose in issuing this volume at the present time is twofold: on the one hand, it is intended to stand for what is known of the flora of the region under discussion; and on the other, by presenting these facts, to incite students of botany to further observation. The publication of a list of species which is still incomplete might be condemned were it not that this is the first published list for the area covered. In studying the flora of a particular section, one of the first needs of the botanist is to know what has been done by others in his field, and, to him, an incomplete list is far better than none.

This work is part of a general plan for an extended study of the flora about the headwaters of the Susquehanna river. There is still much work to be done before our list will approach anything like completeness, and the coöperation of all botanists in the region is desired. It is intended to record all subsequent observations in a series of annual appendices to this volume, and additions and corrections will be gratefully received and proper credit given. To this end, the author offers to identify any specimens which may be sent to him at Binghamton.

In the compilation of the list the fact has been developed that nothing has been published on the flora of our region except a few stray notes in various botanical journals. We have been fortunate, however, in that several able botanists have made our territory the scene of their labors at different times, and this list is based largely upon their notes. Mr. Frederick V. Coville, Botanist of the United States Department of Agriculture, has made a very careful examination of the botany of central Chenango county. Dr. C. F. Millspaugh, now Curator of the Field Columbian Museum, Chicago, made many observations upon the flora in the vicinity of Binghamton and Waverly during ten years' practice of medicine in these places. It was in this region that most of the plants figured in his "American Medicinal Plants" were studied and depicted. Mr. David F. Hoy, Registrar of Cornell University, has collected several seasons in the Susquehanna valley in Delaware county. Dr. Thomas F. Lucy has devoted much of his time to the botany of the Chemung valley for the past thirty years. Mr. James A. Graves has botanized extensively in the lower Chemung valley, and for the past twenty years has made careful observations on the flora in the vicinity of Susquehanna borough, his home. Prof. Frank E. Fenno in the past three years has been indefatigable in investigating the botany of Barton, Apalachin and intermediate points. Mr. W. C. Barbour has contributed many notes on the plants of Sayre, Athens and Waverly, and Miss S. A. Brown has favored us with a very complete list of plants with notes from the vicinity of Unadilla Forks. The author's knowledge of the flora was acquired during a ten years' residence at Binghamton, in which time the greater part of the region came under his observation.

We are fortunate also in having had the species of critical groups passed upon by acknowledged authorities. Doubtful carices have been named by Prof. T. C. Porter and Dr. Charles H. Peck; grasses difficult to identify have received attention from Prof. F. Lamson Scribner and Mr. George V. Nash. The majority of our willows were identified for Dr. Millspaugh by the late M. S. Bebb. Numerous single species of other groups have been named by well-known botanists, who have been given credit in the list. Specimens of a large number of the more critical species have been deposited in the herbarium of Lafayette College, the herbarium of Columbia University, the herbarium of the Field Columbian Museum, and the herbarium of the New York Botanical Garden. Dr. Lucy has also presented to the Elmira Academy of Sciences a nearly complete collection of the plants of our region.

The author is especially indebted to Messrs. Coville, Millspaugh, Hoy, Lucy, Graves, Fenno, Barbour and Miss Brown for not only furnishing copious notes, but also for reading the proof of the entire list, thus authenticating every reference to their own locality. Thanks are also due to Mr. O. S. Wadleigh for his careful attention to the mechanical details of the volume while in press.

WILLARD N. CLUTE.

New York, July 1, 1898.

INTRODUCTION.

THE area drained by the Upper Susquehanna and its tributaries forms an irregular stretch of country about two hundred miles long and from one-third to one-quarter as wide, lying along the forty-second parallel of north latitude, in the States of New York and Pennsylvania. The greater part of this region is in southern New York, and comprises the counties of Otsego, Chenango, Cortland, Broome, Tioga, Chemung and Steuben, with portions of Delaware, Madison, Schuyler and Allegany. In northern Pennsylvania it includes parts of the counties of Wayne, Susquehanna, Bradford and Tioga. On the north, a comparatively low ridge turns the drainage towards the lakes of central New York, and on the south a higher elevation deflects the waters into tributaries of the Susquehanna, which join it much beyond our limits. Upon the east is the drainage system of the Delaware river, and westward the great lakes and the tributaries of the Ohio river receive the rainfall. The flora of this region is of much interest, not alone from the fact that it has only been known in a general way, but also because it is the connecting link between the "Flora of the Lackawanna and Wyoming Valleys" and the "Cayuga Flora," and thus completes a chain of Local Floras extending from eastern Pennsylvania to Lake Ontario.

Topographically, this area may be divided into three sections which are in a measure different from one another. In the extreme east the surface is mountainous, with many steep acclivities, whose summits rise to from fifteen hundred to two thousand feet above tide. The drainage is principally southward by way of the Susquehanna and Chenango rivers. The elevation gradually decreases as the central region is approached. This is occupied by the broad main valley of the Susquehanna extending nearly east and west, with a few short lateral valleys at nearly right angles to it. At this point our region is narrowest, although the surrounding hills seldom rise to heights of more than twelve hundred feet above tide. In the western section the drainage from the north, west and south converges and flows southeastward through a narrow valley as the Chemung river, emptying into the Susquehanna a short distance south of the New York state line. In this part the country again becomes elevated, especially in the south, where the crests rise to heights of two thousand feet or more above tide. After being joined by the Chemung, the Susquehanna enters a deep and narrow canyon, through which it flows for many miles in a tortuous course, and so passes beyond our limits.

GEOLOGY.

Little attention seems to have been paid to the geology of this region; perhaps from the fact that it is considered rather dull. The rocks consist entirely of the Catskill and Chemung formations. Of these the Catskill group is composed of compact sandstones which cap the higher elevations, being most noticeable in the south, east and southwest, and often wanting in the northern and central portions. The valleys of most of our rivers and streams are cut into the softer shales of the Chemung group, which underlie the whole region; in fact the outcrop of these rocks along the Chemung river has given the name to the series. Owing to this softness of the bed rock, our region presents few cliffs and bold precipices, the valleys for the most part sloping gently to the surrounding hills. The flora, therefore, lacks many of the characteristic plants of a more broken country. The greatest exposure of rock is seen in the east and southwest, where the Catskill series predominates. In other parts, the ravines cut by the streams in their descent give us sections of the Chemung shales to the depth of a hundred feet or more.

During the glacial period this district was deeply covered by the ice-sheet. Evidences of this are to be seen in the drift that everywhere strews the surface, and which in the larger valleys lies heaped in ridges, terraces and mounds composed of angular and rounded fragments of shale, limestone, sandstone and quartzite. In its course through southern New York, the Susquehanna flows over a buried valley filled with drift to unknown depths. Large amounts of clay, also of glacial origin, occur in various places.

The soil of the river bottoms is mainly alluvial; on the slopes and uplands it is a clay or gravelly loam derived from the glacial debris, or frequently from the decomposition of the underlying rocks. Some sandy loam is also found, but stretches of pure sand are rare. There is considerable evidence to show that upon the retreat of the ice-sheet a large share of the water from the melting ice was carried off through our region. Older and higher river terraces are still plainly marked along the Chenango, Susquehanna and Chemung, and the deep canyon through which the Susquehanna leaves our region is said to owe much of its depth to these floods.

RIVERS AND STREAMS.

The Susquehanna river rises in Otsego lake, in the northeastern part of this region, and flows in a southwestern direction until shortly after it crosses the line into Pennsylvania. Here it makes the "great bend" west and then northward, back into the State of New York, where it again flows westward for about fifty miles, a short distance from, and roughly parallel to the state line, making its second and final entrance into the State of Pennsylvania a short distance below Waverly. Among its tributaries mentioned in this volume may be noted the following: In Otsego county, the *Unadilla river* flows southward, emptying into the Susquehanna near Sidney, the *Ouleout creek* entering the Susquehanna at Unadilla. In Wayne county, *Starrucca creek* rises, and flowing northwestward through Susquehanna county empties at Lanesboro. In Susquehanna county, *Canawacta creek* and *Drinker creek* flow northward and empty at Susquehanna borough; *Snake creek* flows north and empties at Kirkwood in Broome county; *Choconut creek* flows north and empties at Vestal in Broome county; *Apalachin creek* flows north and empties at Apalachin in Tioga county (N. Y.). In Broome county, the *Little Choconut creek* flows south, emptying at East Union. In Tioga county (N. Y.), *Owego creek* flows south, emptying at Owego. *Cayuta creek*, the outlet of Cayuta lake, flows southeast through Chemung and Tioga counties, emptying at Sayre in Bradford county.

The first important river to join the Susquehanna is the *Chenango*, which flows southwestward through Chenango and Broome counties, emptying at Binghamton. Its principal tributary, the west branch, or *Tioughnioga*, flows southeast through Cortland county and joins it at Chenango Forks, eleven miles from its mouth. The *Otselic river* is the principal tributary of the Tioughnioga. It flows southwestward through Chenango county and empties at Whitney Point, twelve miles above Chenango Forks.

The Susquehanna's western tributary, the *Chemung*, is formed by the union of the Tioga, Conhocton and Canisteo, about forty miles from its mouth. The *Tioga* flows north through Tioga county (Pa), and joins the Chemung at Painted Post in Steuben county. Its first tributary is the *Cowanesque river*, which flows nearly east through Tioga county (Pa.), and empties near Lawrenceville. The *Canisteo river* flows east through Steuben county and empties into the Tioga near Painted Post. The *Conhocton* flows southeast through Steuben county and empties into the Chemung at Painted Post. Of the Chemung's tributaries, *Baldwin creek* flows south through Chemung county, emptying at Lowman P. O. ; *Seeley creek* flows northeast through Chemung county and empties three miles east of Elmira. *Bently creek* rises in Bradford county and flows north, emptying at Wellsburg in Chemung county. *Post creek* flows south in Chemung county, emptying at Corning. *Sing Sing creek* flows south through Chemung county, emptying at Big Flats. *Newtown creek* flows south through Chemung county, emptying at Elmira.

The majority of these are rapid streams, the course of the smaller ones being often interrupted by cascades. In the main valley of the Susquehanna the rate of fall, taken from the railway levels, which closely follow the river, is as follows:

Station.	Above Tide.
Susquehanna	914 feet.
Binghamton	868 "
Owego	822 "
Smithboro	799 "
Tunkhannock	610 "

This would make the fall from Susquehanna to Waverly about two feet to the mile. In this part of its course the river has many deep and quiet coves whose semi-stagnant water forms congenial homes for *Heteranthera dubia, Nymphæa advena, N. microphylla, Valisneria spiralis, Udora Canadensis, Ceratophyllum demersum*, and many pond-weeds. The muddy shores are bordered with *Sparganiums, Alisma Plantago-aquatica, Sagittarias, Lobelia cardinalis*, and various sedges.

LAKES AND PONDS.

Although our region is surrounded on all sides by a multitude of lakes, of glacial origin and otherwise, there are very few within its limits. The most noteworthy in Delaware county are *Goodrich, Mud,* and *Sexsmith lakes,* drained by small tributaries of the Susquehanna. In Susquehanna county, *Comfort's pond* and *Churchill's lake,* drained by Canawacta creek; *Fox's pond* drained by Drinker creek; *Quaker lake,* drained by Snake creek. In Chenango county, *Brisben pond, Warn's pond* and *Geneganslet lake,* drained by branches of the Chenango. In Broome county, *Pond Brook* and *Cutler's pond,* drained by branches of the Chenango. In Tioga county (N. Y.), *Mutton-Hill pond* and *Pemberton's pond,* drained by branches of the Susquehanna. In Schuyler county, *Cayuta lake,* drained by Cayuta creek. In Chemung county, *Miller's pond,* drained by a small tributary of the Chemung. In Steuben county, *Cinnamon lake,* drained by Post creek. In Chenango county, *Round pond* near Greene, *Round pond* near McDonough, and *Jam pond* near Pharsalia, all drained into the Chenango river.

In general the shores of our lakes are composed of soft, yielding mud, pointing unmistakably to the fact that they were once of much greater area and are fast filling up. This also gives a clue to the origin of the numerous peat-bogs in their vicinity, which doubtless were formerly lakes of similar character. In themselves the lake borders are veritable bogs, overgrown with sphagnum and cranberry vines, and tenanted by *Limodorum tuberosum, Pogonia ophioglossoides, Chamædaphne calyculata, Sarracenia purpurea, Drosera rotundifolia, Kalmia glauca* and *Naumburgia thyrsiflora.* As the soil approaches the stable land a home is found for *Alnus, Rhus Vernix, Vaccinium corymbosum, Decodon verticillatus* and *Andromeda Polifolia.* In the lakes a profuse vegetation usually exists, consisting of *Brasenia purpurea, Castalia odorata, Utricularia vulgaris, Nymphæa advena* and others.

BOGS AND SWAMPS.

No part of our territory forms more interesting ground for the botanist than the peat-bogs and swamps. Of from one to many acres in area, they range from comparatively solid peat to "quaking bogs" of unfathomed depth, over whose surface even the botanist may not pass. They are usually covered with sphagnum and afford a home for plants similar to those that frequent the lake borders. Several, although true bogs, are called swamps. Among the more interesting may be mentioned *Thompson's marsh*, a short distance east of Pond Brook; the *Cranberry marsh*, two miles north of Lanesboro; the *peat-bog* near Jarvis street in the city of Binghamton; *Beebe's swamp*, along the Susquehanna at Oakland; *Bear swamp*, one mile southeast of Susquehanna and several hundred feet above the river; the *peat-bog* one mile east of Union, in the river valley; *Drake's swamp*, four miles north of Barton and nearly four hundred feet above the river; *Tribe's swamp*, two miles north of Drake's swamp; *VanEtten swamp*, sixteen miles north of Sayre between VanEtten and Spencer; the *Vlai*, on the ridge between the Ouleout and Susquehanna, about a mile and a half from Oneonta, and the *Brisben bogs*, about half-way between Oxford and Greene and a mile and a half from Brisben.

Second only to the bogs are the swamps proper. These are of less depth, contain little or no peat and are usually covered with several inches of standing water. They are the homes of *Typha latifolia, Calla palustris, Nymphæa advena, Ilex verticillata, Sparganium*, and a large number of sedges. Of this class are the *Beechwood swamp* at the west end of Clinton street, Binghamton, the *marshlands*, a short distance west of Apalachin, near the river, and *Lowman's swamp*, at Lowman P. O.

MOUNTAINS AND RAVINES.

There are few if any elevations within our limits of sufficient altitude to be called mountains, but several lesser heights have been dignified by the title. Among them, *Mt. Prospect*, northwest of Binghamton, *Ely hill*, northeast of Binghamton, and *South Mountain*, south of Binghamton, have elevations of about 1,200 feet above tide. *Mutton-Hill*, south of Apalachin, elevation 1,350 feet; *Spanish Hill*, two miles northwest of Sayre, elevation about 900 feet; *Sullivan Hill*, five miles east of Elmira, elevation about 1,500 feet. The *Chemung Narrows*, west of Chemung village, *Mountain House Narrows*, five miles west of Elmira, and the *Narrows at Owego*, are so called from the fact that here steep cliffs constrict the valley.

Nearly all the smaller streams make their way to the lowlands through ravines which are often fifty feet or more in depth. Their sides are usually quite steep, but afford a foothold for a large number of species that love the shade, especially ferns. The ravines are all particularly rich in species. Of those mentioned in the following pages, *Canavan's Glen* is west of Susquehanna; *Pope's ravine*, northeast of Binghamton; *Glenwood ravine*, northwest of Binghamton, and *Roericke's Glen*, two miles above Elmira.

ALTITUDE OF PRINCIPAL POINTS.

The principal stations mentioned in the list of species are given below, with their elevation above tide, taken from various railway levels.

STATION.	RAILWAY LEVEL.	ABOVE TIDE.
Brandts	Erie	1047
Ararat Summit	"	2023
Lanesboro Junction	"	983
Susquehanna	"	914
Great Bend	"	884
Binghamton	"	869
Union	"	834
Owego	"	822
Waverly	"	836
"	Lehigh Vall	830*
Sayre	" "	773
Towanda	" "	737†
Tunkhannock	" "	610
Wellsboro	Erie	831
Elmira	"	863
Corning	"	942
Painted Post	"	947
New Milford	D., L. & W.	1087
Vestal	" "	826
Apalachin	" "	819
Unadilla Forks	D. & H.	1194
Norwich	D., L. & W.	1014
Oxford	" "	980
Chenango Forks	" "	960

* Track stated to be 80 feet above Chemung river.
† Track stated to be 38 feet above Susquehanna river.

TEMPERATURE AND RAINFALL.

The subjoined tables of the temperature and rainfall over our region have been furnished by the United States Weather Bureau. Although not within our limits, Ithaca has been included for purposes of comparison.

TABLE OF MEAN TEMPERATURE (Degrees Fahr.).

	Jan.	Feb.	Mar.	Apr.	May	June	July	Aug.	Sept.	Oct.	Nov.	Dec.	Annual	
Binghamton...	22.2	23.6	30.2	45.5	57.0	67.2	68.2	67.3	61.1	47.7	38.0	28.3	46.5	a
Elmira*......	27.3	27.0	33.2	48.6	60.2	70.0	72.4	70.3	63.7	49.3	40.6†	32.5†	49.6	b
Waverly	23.2	24.5	30.1	44.9	57.7	66.9	69.1	67.0	60.4	47.8	37.9	29.0	46.5	c
Ithaca.........	23.1	24.7	29.3	43.4	56.9	66.2	69.9	67.2	60.8	48.5	38.7	29.4	46.6	d

    a Means for 1891 to 1896 inclusive. Also includes July to Dec., inclusive, for 1890.
    b For 1889 to 1896, inclusive.
    c From 1883 to 1896, inclusive. Also includes July to Dec., inclusive, for 1882.
    d From 1881 to 1890, inclusive.
    * The high temperature at this station may be due to local causes.
    † November and December means missing for 1892-'93.

TABLE OF PRECIPITATION (In Inches and Hundredths).

	Jan.	Feb.	Mar.	Apr.	May	June	July	Aug.	Sept.	Oct.	Nov.	Dec.	Annual	
Binghamton...	3.06	3.03	3.17	2.18	3.96	3.28	3.34	4.30	3.30	3.18	2.25	2.67	37.72	a
Elmira.........	2.11	1.82	1.92	2.41	4.50	3.86	3.18	3.30	3.26	3.18	1.80	2.45	33.79	b
Waverly.......	2.33	2.05	2.24	2.14	3.63	3.39	3.61	3.26	3.29	3.10	2.34	2.23	33.50	c
Ithaca.........	2.28	2.00	4.23	1.98	3.88	3.81	3.85	3.45	2.76	3.40	2.59	2.41	34.65	d

    a Means for 1891 to 1896, inclusive. Also includes means for June to December, 1890, inclusive.
    b Means from 1889 to 1896, inclusive.
    c Means for 1882 to 1896, inclusive.
    d Means for fourteen years.

By a reference to the map, it will be seen that a line connecting the farthest points from which this data has been recorded will form an isosceles triangle whose longest side is in the main valley and very near the centre of our region. From the conditions existing within this triangle we may form a fair idea of those that prevail over the rest, though it is to be regretted that records from other parts are not obtainable.

It is generally understood that the return of the birds, the blooming of the first flowers and the general revival of animal life in the spring, bears a certain relation to the temperature of a

ARRIVAL OF BIRDS AT BINGHAMTON.

	1886	1887	1888	1889	1890	1891	1892	1893	1894	1895
Bluebird	Mar. 15	Mar. 12	Feb. 26	Feb. 22	Feb. 24	Feb. 23	Feb. 20	Mar. 8	Mar. 4	Mar. 22
Robin	" 18	" 12	Mar. 21	Mar. 12	Mar. 9	Mar. 12	" 27	" 12	" 5	" 25
Song Sparrow	" 18	" 12	" 24	" 13	" 13	" 17	Mar. 14	" 14	" 7	" 25
Crow Blackbird	" 28	Apr. 3	" 28	Apr. 6	" 30	" 23	" 28	" 14	Apr. 7	" 25
Red-winged Bl'kbird	" 28	" 3	Apr. 28	" 18	Apr. 3	Apr. 3	Apr. 2	" 14	" 20	Apr. 12
Meadow Lark	Apr. 6	" 11	" 31	" 17	" 30	" 12	" 2	" 26	Mar. 18	Mar. 30
Pewee	" 2	" 14	" 31	" 3	Apr. 4	" 14	" 3	" 28	Apr. 19	Apr. 5
Flicker	.....	Apr. 14	Apr. 3	" 11	" 5	" 14	" 5	Apr. 9	Mar. 11	" 24
Chipping Sparrow	.....	" 11	" 7	" 11	" 12	" 15	" 2	" 14	Apr. 20	" 26
Purple Finch	Apr. 9	" 10	" 11	" 4	" 13	" 8	.....	Mar. 31	May 17	" 28
Junco	.....	Mar. 30	Mar. 24	" 6	Mar. 30	.....	.....	Apr. 19	Mar. 30	" 11
White-bellied Swal'w	.....	Apr. 11	Apr. 17	" 12	Apr. 8	" 13	Apr. 4	" 16	Apr. 10	" 26
Chimney Swift	May 4	May 2	" 28	" 19	May 2	" 19	" 29	May 1	May 2	May 1
Oriole	" 6	" 3	May 4	May 8	" 2	May 2	May 6	" 13	" 5	" 5
Kingbird	.....	" 5	" 20	" 11	" 16	" 3	.....	.....	" 12	" 18

BLOOMING OF FIRST FLOWERS AT BINGHAMTON.

	1891	1892	1893	1894	1895
Hepatica Hepatica	Mar. 31	Apr. 3	Apr. 1	Mar. 18	Apr. 11
Epigæa repens	Apr. 17	" 15	" 23	Apr. 10	" 18
Syndesmon thalictrioides	" 15	" 27	" 29	May 2	" 24
Taraxicum Taraxicum	" 18	" 17	" 30	Apr. 17	" 28
Erythronium Americanum	" 18	" 26	May 7	" 19	" 25
Houstonia cerulea	" 19	" 24	" 9	Mar. 11	" 14
Viola blanda	" 26	May 30	" 7	Apr. 20	" 28
Caltha palustris	" 19	May 8	" 10	May 1	" 30
Saxifraga Virginiensis	" 26	Apr. 23	" 15	Apr. 27	" 24
Anemone quinquefolia	May 2	May 1	" 7	" 22	" 30
Mitella diphylla	" 3	" 5	" 15	May 2	May 5
Trillium grandiflorum	Apr. 30	" 8	" 9	" 2	" 1

	1887	1889	1890	1891	1892	1893
First Frog seen	Apr. 15	Mar. 22	....	Apr. 6	Apr. 2	Mar. 23
" Hyla heard	....	Apr. 7	....	" 6	" 2	....
" Butterfly seen	Apr. 3	Mar. 23	Apr. 5	" 18	" 3	Mar. 29

region. The preceding tables, compiled from records made at Binghamton, may be of interest. In making comparisons it should be remembered that Binghamton is situated in a broad valley, surrounded by low hills, and at an elevation of 868 feet above tide. In some parts of our territory the seasons are nearly two weeks later than here.

CHARACTERISTICS OF THE FLORA.

A more careful examination of the remote districts which have been but slightly studied, will doubtless result in the discovery of a considerable number of species new to our list. Until this has been done, it will be impossible to make a thorough comparison of the flora with that of other regions, but the present list is sufficient to indicate the general relationships, which subsequent additions of any number of species are not expected to materially change.

NORTHERN PLANTS. The influence of a boreal flora is seen in the number of plants common to higher latitudes which here nearly reach their southern limits. If found farther south they generally occur in elevated districts, or in cold bogs and ravines. Among these may be mentioned *Oxalis Acetosella, Comarum palustre, Linnæa borealis, Schollera Oxycoccus, Chiogenes hispidula, Andromeda Polifolia, Kalmia glauca, Rhodora Canadensis, Ledum Grœnlandicum, Moneses uniflora, Menyanthes trifoliata, Betula papyrifera, Larix Americana, Calla palustris, Carex pauciflora* and *Botrychium simplex.*

SOUTHERN SPECIES. That our flora has less relationship with that of the south, is shown by the following list of plants that occur, most of which find in our valley their farthest northern range: *Magnolia acuminata, Jeffersonia diphylla, Acer Negundo, Robinia pseudacacia, Hydrangea arborescens, Azalea canescens, Rhododendron maximum, Cunila origanoides, Plantago Virginica, Juglans nigra, Carya sulcata, Dioscorea villosa, Disporum lanuginosum, Eragrostis Caroliniana* and *E. Frankii.* One western species, *Kœleria cristata,* finds its northeastern limits with us.

RARE PLANTS. There are certain plants which, though they may be widely disseminated, are usually rare in any region. That ours is not lacking in these is shown by the occurrence of such as *Atragene Americana, Aconitum Novcboracensis, Cacalia suaveolens, Razoumofskya pusilla, Pogonia verticillata, Carex Schweinitzii, C. abacta* and *Pellæa gracilis.* Of plants that are unusual in our region may be mentioned *Cardamine pratensis, Ptelea trifolia, Staphylea trifoliata, Celtis occidentalis, Woodwardia Virginica, Lygodium palmatum* and *Pterospora Andromedea.*

INTRODUCED PLANTS. Aside from the regular list of introduced species which have made their way into many parts of America and become so much at home that they appear as if native, our region furnishes some of so recent introduction that the process of naturalization may be said to be still going on. Good examples of this are found in *Trifolium hybridum, Hesperis matronalis, Scabiosa australis, Hieracium aurantiacum, H. pilosella, Tragopogon pratensis, Cichorium Intybus, Lactuca Scariola, Cuscuta Epithymum, Echium vulgare, Solanum rostratum, Euphorbia Niceænsis, E. peplus* and *Iris pseudacorus.*

THE LESSER FLORAS.

Not the least interesting feature of our region is found in the natural grouping of plants into lesser floras. Peculiar conditions of soil, temperature or topography serve to limit these areas and in many cases the lines are decidedly marked. Some of the more important are given herewith

POND BROOK. This name is applied to two small lakes lying in the valley of the Chenango, about two miles south of its junction with the Tioughnioga. In a general way it is also applied to the whole of the surrounding region, whose surface is here most peculiar, consisting of a series of great depressions separated from one another by low, steep banks of glacial debris. The region probably dates its present formation from the retreat of the ice-sheet. Two of the depressions hold the lakes, whose bottoms are scarcely above the level of the river, and the rest contain peat-bogs overgrown with sphagnum. This region is probably richer in species than any other place of like extent within our limits. Not to mention commoner things, it contains *Drosera rotundifolia, Sarracenia purpurea, Nymphæa advena, Decodon verticillatus, Cardamine pratensis, Castalia odorata, Naumburgia thyrsiflora, Cladium mariscoides, Utricularia vulgaris, Batrachium trichophyllum, Rynchospora alba, Eriophorum polystachyon, Scheuchzeria palustris, Cinna arundinacea, Chiogenes hispidula, Azalea canescens, A. nudiflora, Lobelia Kalmii, Cypripedium hirsutum, C. acaule, Comarum palustre, Atragene*

Americana, Menyanthes trifoliata, Schollera Oxycoccus, Clintonia borealis, Chamædaphne calyculata, Xolisma ligustrina, Andromeda Polifolia and *Dryopteris Boottii*. The dividing ridges are wooded with a variety of deciduous trees and add their quota of rarities.

THOMPSON'S MARSH. This is but a continuation of Pond Brook lying over the next ridge, but the flora differs enough from the other to merit a word. In addition to a majority of the species presented by the former, it yields *Orontium aquaticum, Aralia hispida, Pogonia ophioglossoides, Woodwardia Virginica*, and is fringed by immense thickets of *Aronia nigra* and *Vaccinium corymbosum*.

MUTTON-HILL POND. This pond, containing about twenty acres, lies in a hollow in the hilltops, two miles west of Apalachin and several hundred feet above the river. It is surrounded by a broad belt of shaking peat-bog that will scarcely sustain the weight of a man, Among the species peculiar to it are *Nymphæa advena, Castalia odorata, Menyanthes trifoliata, Pogonia ophioglossoides, Schollera Oxycoccus, Comarum palustre, Drosera rotundifolia, Calla palustris, Eriophorum Virginianum, E. polystachyon, Rynchospora alba, Ilicioides mucronata, Ilex verticillata, Rhus Vernix* and *Xolisma ligustrina*.

BOG NEAR JARVIS STREET. This bog of less than five acres in extent, lies at the north end of Jarvis street in the city of Binghamton. It is filled with peat to a considerable depth and at most seasons the remnant of the lake that once occupied the site is still visible near the center. Although so exposed and within the city, it yields a large number of the less common species, such as *Chamædaphne calyculata, Andromeda Polifolia, Schollera Oxycoccus, Orontium aquaticum, Hypericum Virginicum, Nymphæa advena, Rosa Carolina, Comarum palustre, Pogonia ophioglossoides, Acorus Calamus* and *Cephalanthus occidentalis*.

BEEBE'S SWAMP. This marsh, originally called "Cranberry marsh," from the abundance of that berry once growing there, lies in a deep basin some distance back from the river and opposite the borough of Susquehanna. Its flora is of interest from the number of distinctive species that occur. Of the more noticeable may be mentioned *Gaylussacia resinosa, Schollera Oxycoccus, S. macrocarpa, Andromeda Polifolia, Xolisma ligustrina, Chamædaphne calyculata, Kalmia angustifolia, K. glauca, Rhodora Can-*

adensis, *Cephalanthus occidentalis*, *Cornus Amomum*, *C. stolonifera*, *Rosa Carolina*, *Ilicioides mucronata*, *Ilex lævigata*, *Aronia nigra*, *Hypericum Virginicum*, *Ranunculus obtusiusculus*, *Sarracenia purpurea*, *Carex abacta*, *C. Schweinitzii*, *C. bullata*, *Dryopteris spinulosa* and *Woodwardia Virginica*.

GLENWOOD RAVINE. Our ravines all possess floras of similar character. The shade and moisture make congenial homes for ferns, while a great number of flowering plants are also to be found. Among the plants of Glenwood ravine may be mentioned *Viola rotundifolia*, *Polygonatum biflorum*, *Uvularia grandiflora*, *U. sessilifolia*, *U. perfoliata*, *Arisæma triphyllum*, *Viburnum alnifolium*, *Acer spicatum*, *A. Pennsylvanicum*, *Viola blanda* and *V. pubescens*. On the high lands surrounding it may be found *Epigæa repens*, *Hypoxis erecta*, *Kalmia latifolia*, *Pogonia verticillata*, *Cypripedium hirsutum*, *C. parviflorum*. *Lupinus perennis*, *Azalea nudiflora*, *Lilium Philadelphicum*, *Ranunculus fascicularis*, *Dasystoma Virginica* and *Cornus florida*.

THE MARSHLAND. This bit of marsh, although not differing much from the surrounding country in external appearance, has a flora decidedly peculiar to it. Among its species occur *Comarum palustre*, *Nymphæa microphylla*, *Aronia nigra*, *Ilex verticillata*, *Ilicioides mucronata*, *Panicularia aquatica*, *Alopecurus geniculatus*, *Chamædaphne calyculata* and *Cephalanthus occidentalis*.

THE RIVER BANKS. That part of the river which extends east and west through New York has not only a flora peculiar to it, but one that differs in different parts, though seemingly exposed to to the same natural conditions. Among the species that give character to the banks are *Sanguinaria Canadensis*, *Dentaria diphylla*, *D. laciniata*, *Trillium erectum*, *T. grandiflorum*, *Erythronium Americanum*, *Mertensia Virginica*, *Lobelia cardinalis*, *Gaura biennis*, *Impatiens aurea*, *I. biflora*, *Apocynum cannabinum*, *Salix nigra*, *Claytonia Virginica*, *Arisæma Dracontium*, *A. triphyllum*, *Ambrosia trifida*, *Micrampelis lobata*, *Sicyos angulatus* and *Onoclea sensibilis*. These are generally distributed throughout the valley, but there are other plants with a more curious distribution. Among these, *Celtis occidentalis*, *Asclepias tuberosa*, *Menispermum Canadense*, *Staphylea trifoliata*, *Dioscorea villosa*, *Dianthera Americana* and *Leptandra Virginica*, may be cited. At Barton all these are common, but as one travels eastward, each in its turn disappears. At Smithboro the last of *Dianthera Americana* is seen; *Staphylea trifoliata* and *Asclepias tuberosa* become rare at Apalachin and disappear at Binghamton. *Menispermum Canadense* and *Dioscorea villosa* are not found much beyond Binghamton; *Celtis occidentalis* is common as far as Apalachin, and becoming rare, disappears at Binghamton, as does also *Leptandra Virginica*. Several others show this peculiarity of distribution in various degrees.

STATISTICS OF THE FLORA.

Although the list of species includes all that are at present known to grow without cultivation within our limits, a comparison of the flora with others upon the mere basis of numbers would be unfair at this time. In our efforts to make this first published list of the plants of our region thoroughly authentic, many plants that doubtless should have had a place have been excluded. The practice has been to throw out every species about which there might be the slightest doubt. It is gratifying, however, to report more than a thousand species. The position our flora holds in relation to others with regard to numbers is shown by the following table:

```
Plants of Monroe county, etc., N. Y........1,314
Cayuga Flora............................1,278
Plants of Buffalo and vicinity. ..........1,243
Flora of Wyoming and Lackawanna valleys.1,107
Flora of the Upper Susquehanna. .........1,105
Plants of Dutchess county, N. Y..........1,067
Plants of Suffolk county, N. Y............  852
```

A comparison of our list of species with the larger lists will show that had we included the common escapes of the garden and counted a considerable number of native varieties, as has been done with them, the difference in numbers would be slight, if any.

NOMENCLATURE OF THE LIST.

The nomenclature of the list throughout is that of the "List of the Pteridophyta and Spermatophyta growing without cultivation in Northeastern America," published as Volume V. of the "Memoirs of the Torrey Botanical Club." When these names differ from the names in the sixth edition of "Gray's Manual," the names of the latter follow in parenthesis the remarks on the species.

It may be said of the names used that they are essentially those of the "new nomenclature." Our aim has been to designate exactly the species in mind without attempting to follow the various changes that have been made since the list of the Torrey Club was published. When nomenclature becomes more stable, this list will be placed in accord with it in an appendix. In the arrangement of genera and species we have chosen to follow that of the sixth edition of "Gray's Manual," as it is the one most familiar to those who will have use for this book.

When more than one common name is given to a species, the name that is most commonly used in our region is placed first. The relative abundance of the species follows the common name, and when this is qualified further along in the citation it is to be taken to refer only to the clause with which the qualification is connected. The other observations are designed mainly to aid young botanists to a knowledge of some of the interesting facts about the species which are frequently missing from the text-books.

The authority for the statements in the list is indicated by the surnames of the observers in italic. When no locality is given with the observation, it should be understood to refer to localities as follows: *Coville*, central and northern Chenango county; *Lucy*, valley of the Chemung; *Graves*, Susquehanna county; *Fenno*, Apalachin to Barton; *Brown*, vicinity of Unadilla Forks; *Barbour*, Sayre and vicinity; *Hoy*, Delaware county; *Clute*, Broome county.

LIST OF SPECIES.

RANUNCULACEÆ.

CLEMATIS L.

C. Virginiana L. WHITE CLEMATIS. VIRGIN'S BOWER. TRAVELER'S JOY. A common vine about fence rows, river banks and thickets. Most abundant in damp situations. Often mistaken for poison ivy. The globular, feathery fruiting heads are well known in autumn. July.

ATRAGENE L.

A. Americana Sims. PURPLE CLEMATIS. MOUNTAIN CLEMATIS. Rare. Found in rocky situations. Hills above Susquehanna, *Graves*. Pond Brook, *Clute*. South Mountain, *Millspaugh*. Southport, Corning, *Lucy*. May. (*Clematis verticillaris* DC.)

ANEMONE L.

A. Virginiana L. VIRGINIAN ANEMONE or WIND-FLOWER. THIMBLE WEED. Common along river banks and the borders of woods. June. July.

A. quinquefolia L. WIND-FLOWER. WOOD ANEMONE. NIGHTCAPS. Common in thickets along streams and in moist woodlands. The flower is colored outside with purple, which varies considerably in depth of tint in different individuals. May. (*A. nemorosa* L.)

A. cylindrica A. Gray. LONG-FRUITED ANEMONE. Reported only from the eastern part of our range. Along roads; scarce, *Graves*. Frequent along the railroad, Sidney, *Hoy*. Borders of fields, Vestal, *Millspaugh*. May.

A. Canadensis L. PENNSYLVANIAN ANEMONE. Common in moist soil. Not reported from the Chemung valley. June–Aug. (*A. Pennsylvanica* L.)

HEPATICA L.

H. Hepatica (L.) Karst. LIVERLEAF. LIVER-WORT. SQUIRREL-CUPS. HEPATICA. Found throughout. Abundant in thickets and deciduous woods from the Chemung valley east to the rugged parts of Susquehanna and Delaware counties, where it is in part displaced by the next species. A very scarce plant, *Coville*. Not common at Sidney and Franklin, *Hoy*. It is easily cultivated. One of our earliest plants to bloom. Flowers of various shades of white, pink and blue; occasionally fragrant. Sometimes blooms in autumn. March–June. (*H. triloba* Chaix.)

H. acuta (Pursh) Britton. ACUTE-LOBED HEPATICA or LIVERLEAF. Occurs in the same situations as the preceding, but seeming to prefer more cold and moisture. It gradually runs out as the Chemung valley is approached. Less frequent than the last, *Lucy*. Barton, *Fenno*. In a cold ravine, Choconut Centre; Killawog, *Clute*. About as common at the other, *Graves*. Very common at Franklin, *Hoy*. Abundant in the upper Chenango valley, *Coville*. Not reported from Tioga county. The leaves are frequently five-lobed and the flowers apparently less variable in color. April–June. (*H. acutiloba* DC.)

SYNDESMON HOFFMG.

S. thalictrioides (L.) Hoffmg. RUE ANEMONE. Abundant in open woods, thickets and banks along streams. Rare in beech and maple woods, *Fenno*. One of our earliest spring flowers. Easily cultivated. Stem arising from a cluster of tuberous roots. Apr. May. (*Anemonella thalictroides* Spach.)

THALICTRUM L.

T. dioicum L. EARLY MEADOW RUE. Common in shaded, rocky soil. Apr. May.

T. polygamum Muhl. TALL MEADOW RUE. Common in rich, moist soil, especially in wet meadows and along streams. Very noticeable when in bloom. July.

T. purpurascens L. PURPLISH MEADOW RUE. Rare. Found in drier soil than the preceding. Harrington's Ford, Chemung river, *Lucy*. Mt. Prospect, *Millspaugh*. July.

BATRACHIUM S. F. Gray.

B. trichophyllum (Chaix) Bossch. White Water Crowfoot. Found only in the eastern part of our range, in shallow water. Scarce, *Graves*. Common at Davenport and Goodrich Lake, *Hoy*. Chenango river, *Coville*. Hoboken pond, near New Berlin, *Ellsworth*. July. (*R. aquatilis* L., var. *trichophyllus* Gray.)

RANUNCULUS L.

R. delphinifolius Torr. Yellow Water Crowfoot. Rare. Pemberton's Pond, Waverly, *Millspaugh*. Beechwood swamp, *Clute*. Leaves finely dissected, immersed; flowers rather large, bright yellow. May. June. (*R. multifidus* Pursh.)

R. obtusiusculus Raf. Water Plantain Spearwort. Rare. Beebe's swamp, *Graves*. Rockbottom dam, *Millspaugh*. Beechwood swamp, *Clute*. Wellsburg; Painted Post, *Lucy*. Flowers small; leaves lanceolate. At a little distance the casual observer might mistake this plant for a crucifer. July. (*R. ambigens* Wats.)

R. reptans L. Small Creeping Spearwort. Found occasionally in gravelly soil on the margin of the Susquehanna river. Near Apalachin, not common, *Fenno*. West of Binghamton, *Clute*. Stems rooting at the joints; leaves linear; flowers very small. July. (*R. Flammula*, var. *reptans* Meyer.)

R. pusillus Poir. One station only—a springy place at the base of South Mountain, *Millspaugh*.

R. abortivus L. Small Flowered Crowfoot. Common throughout in moist soil, in fields, thickets and the borders of woods. May. June.

R. sceleratus L. Cursed Crowfoot. Rare. Usually found along ditches. Apalachin, *Fenno*. Waverly, *Millspaugh*. Chenango county, *Coville*. New Berlin, *Ellsworth*.

R. recurvatus Poir. Hooked Crowfoot. Common in damp woods. Flowers resembling somewhat those of *R. abortivus*, but larger. May. June.

R. fascicularis Muhl. Early Crowfoot or Buttercup. Fascicled Crowfoot. Common. Found blooming on hilltops and thinly wooded slopes in company with the Hepatica and Rue Anemone. Seldom found in the lowlands. Our earliest Crowfoot. May.

R. septentrionalis Poir. SWAMP BUTTERCUP. Found usually in swamps and along streams. Produces long runners in summer. The plant is quite variable in size and foliage, and in certain phases may be easily mistaken for *R. fascicularis* and *R. repens*. June–August.

R. repens L. CREEPING CROWFOOT or BUTTERCUP. Common. Oxford, *Coville*. Chenango valley near Binghamton, *Millspaugh*. Common along the Susquehanna at Apalachin, *Fenno*. Very common at Sayre and Lockwood, *Barbour*. A variety of this, with white-spotted, hairy leaves and prostrate stems, has found its way into many lawns in the city of Binghamton. The whole plant is so low that mowing does not harm it, *Clute*. July. August.

R. Pennsylvanicus L. BRISTLY CROWFOOT. Found in wet soil, usually in the open. Scarce, *Graves*. Noyes island, *Millspaugh*. Willow Point, not common, *Clute*. Common, *Fenno*. Occasional, *Lucy*. Upper Unadilla valley, *Ellsworth*. July. August.

R. hispidus Michx. Moist woods in Chemung county. Not infrequent, *Lucy*. Has been confused with *R. repens, fascicularis* and *septentrionalis*. Should be looked for in other parts.

R. bulbosus L. BULBOUS CROWFOOT. Found sparingly in pastures near Oxford, *Coville*. Not common along the Unadilla river, *Brown*. South slope Asylum hill, Binghamton, *Millspaugh*. Just south of our range this plant is very common in certain localities.

R. acris L. COMMON BUTTERCUP. TALL BUTTERCUP. Abundant, especially in wet meadows, where it often forms the principal vegetation. June–August.

CALTHA L.

C. palustris L. MARSH-MARIGOLD. COWSLIP. Abundant in swamps and wet woodlands; rarer in the western part of our range. Not common, *Fenno*. Rare in Chemung county, *Lucy*.

North of Chemung county this species is more common. The leaves and blossoms are frequently used as greens in spring. May.

COPTIS Salisb.

C. trifolia (L.) Salisb. GOLD-THREAD. THREE-LEAVED COPTIS. MOUTHROOT. Common, especially in low, wet woods. Found on hummocks and about stumps and old logs. Leaves, evergreen; stems, subterranean, long, thread-like, bright yellow, bitter; petals, club-shaped, hollow at the apex. May.

AQUILEGIA L.

A. Canadensis L. COLUMBINE. HONEYSUCKLE. Common in wet meadows and on rocky banks; seldom in deep woods. An ornamental plant, easily cultivated. May. June.

A. vulgaris L. GARDEN COLUMBINE. Occasionally found as an escape along roadsides.

ACONITUM L.

A. Noveboracense A. Gray. ACONITE. MONKSHOOD. WOLFBANE. Very rare. This species was first described from specimens found on the bank of the Chenango river two miles below Oxford, in 1883. The herbarium of Columbia University contains a specimen collected at Greene, N. Y., by Dr. Augustus Willard about the year 1857, *Coville.* Has since been found in Orange county.

CIMICIFUGA L.

C. racemosa (L.) Nutt. BUGBANE. BLACK SNAKEROOT. BLACK COHOSH. Very common along river banks, in thickets and the borders of woods. Rare at Oxford and northward, *Coville.* Flowers numerous, ill-scented, in long terminal spikes. Very noticeable when in bloom. June. July.

ACTÆA. L.

A. rubra (Ait.) Willd. RED BANE-BERRY. RED COHOSH. Common throughout, especially in rocky woodlands. Very abundant along the Chenango in the vicinity of Lisle, *Clute.* In fruit distinguished from the following species by its slender pedicels and red berries. Flowers ill-scented. May. June. (*A. spicata* L., var. *rubra* Ait.)

A. alba (L.) Mill. WHITE BANE-BERRY. WHITE COHOSH. Common in deciduous woods; apparently more abundant than the preceding in the southern part of our range. Pedicels, thick, red; berries, snow-white. A pink-berried form is occasionally seen, *Graves*. Occasionally white berries have slender pedicels and red berries thick pedicels, but this is supposed to be due to crossing. May.

XANTHORRHIZA L'Her.

X. apiifolia L'Her. SHRUB YELLOWROOT. "In a dark ravine at Sherburne, eleven miles from Hamilton. (*Dr. Douglass.*)"—Torrey's Flora of New York. Apparently has not since been recorded.

MAGNOLIACEÆ

MAGNOLIA L.

M. acuminata L. CUCUMBER TREE. Represented by a few scattered trees. Rare, *Fenno*. Very Scarce, *Lucy*. Not uncommon near Waverly, *Barbour*. Along the Chemung river, 1862, *Graves*.

LIRIODENDRON L.

L. tulipifera L. WHITEWOOD. TULIP TREE. POPLAR. CUCUMBER TREE. Reported as a shade tree from Susquehanna and Chemung counties. It appears to be indigenous, though not common, in the counties of Broome and Tioga. Ross Park, rare, *Clute*. Apalachin; not common, *Fenno*. Mouth of Cayuta creek, *Millspaugh*. Lockwood; not common, *Barbour*. In localities where the Magnolia is unknown this is usually called Cucumber tree. Easily distinguished from our other forest trees by its leaves with truncate middle lobes, and its large green and orange flowers. May. June.

MENISPERMACEÆ.

MENISPERMUM L.

M. Canadense L. MOONSEED. Common from Broome county west. Not reported from Susquehanna and Delaware counties. Common in Chenango county, *Coville*. Restricted to the river shores and banks of the larger creeks, where it climbs over trees and bushes and is frequently mistaken for a grapevine. Berries black; seeds crescent-shaped. June.

BERBERIDACEÆ.

BERBERIS L.

B. vulgaris L. BARBERRY. Not common. Found in old fields, along roads, etc. Local and scarce, *Lucy*. Lockwood; not common, *Barbour*. Mouth of Cayuta creek, *Millspaugh*. River banks in the city of Binghamton, *Clute*. Roadsides, *Graves*. On dry hillsides, *Coville*. Thickets, occasional, *Brown*. New Berlin and western Otsego county, *Ellsworth*. Remarkable for its bristly serrate leaves and irritable stamens, which spring towards the pistil when touched. Cultivated for both ornament and use. The red, acid fruit is often used in jelly-making. June.

CAULOPHYLLUM MICHX.

C. thalictroides (L.) Michx. BLUE COHOSH. PAPPOOSE ROOT. Common throughout in woods and ravines in rich, moist soil. May.

JEFFERSONIA BART.

J. diphylla (L.) Pers. TWIN-LEAF. RHEUMATISM ROOT. Very rare. One station only—a wooded hillside near Dansville, N. Y., *Lucy*.

PODOPHYLLUM L.

P. peltatum L. MAY-APPLE. MANDRAKE. WILD LEMON. RACCOON BERRY. Abundant in woods and thickets, and along the streams throughout our range, growing best in moist places. Well-known from its curious, umbrella-shaped leaves, fragrant flowers and edible fruit. Occasionally two or more fruits are borne on a single stem. May. June.

NYMPHÆACEÆ.

BRASENIA SCHREB.

B. purpurea (Michx.) Casp. WATER SHIELD. WATER TARGET. Common in ponds and lakes. Cayuta Lake, *Dudley*. Mutton-

Hill Pond, *Fenno*. Pond Brook and the coves of the Chenango and Susquehanna, *Clute*. Common in nearly all the ponds in Chenango county, *Coville*. Goodrich Lake, *Hoy*. Common in New Berlin and western Otsego county, *Ellsworth*. Butler's Lake, *Graves*. The latter station contains great quantities of this plant. Leaves elliptical, floating; the petioles attached at the centers; flowers purple; stems of leaves and flowers coated with a jelly-like substance. July. (*B. peltata* Pursh.)

CASTALIA Salisb.

C. odorata (Dryand.) Woodv. & Wood. Water Lily. Water Nymph. Common in quiet waters. Cayuta Lake, *Dudley*. Miller's Pond, *Lucy*. Pemberton's Pond, *Millspaugh*. Mutton-Hill Pond, *Fenno*. Near Spencer, N. Y., *Barbour*. Pond Brook and coves of the Tioughnioga near Lisle, *Clute*. Churchill's Lake, *Graves*. Goodrich Lake, *Hoy*. Silver Lake, New Berlin, *Ellsworth*. Ponds and river coves, *Coville*. Flowers white, occasionally with a tinge of pink, fragrant, open only in the morning. Petals numerous, passing insensibly into stamens. June–Sept. (*Nymphæa odorata* Ait.)

C. tuberosa (Paine) Greene. Tuber-bearing Water Lily. One station—Mud Lake at Davenport, *Hoy*. (*Nymphæa reniformis* DC.)

NYMPHÆA L.

N. advena Soland. Yellow Pond-Lily. Spatterdock. Cow Lily. Frog Lily. Abundant in lakes and other still waters; occasionally in swamps and bogs. In lakes and ponds the leaves float; in coves and inlets of the rivers only the earliest do so; in bogs all may be borne on upright stems. Flowers globular, yellow, sweet-scented, the coarse sepals hiding several rows of scale-like petals. June. July. (*Nuphar advena* Ait. f.)

N. microphylla Pers. Small Yellow Pond-Lily. Kalm's Pond Lily. Rare. Lowman's Cove, *Lucy*. Pond at Franklin, *Hoy*. Cutler's Pond, *Millspaugh*. Susquehanna at Union, Chenango at Port Dickinson, Tioughnioga near Lisle, *Clute*. Pond at Preston, *Coville*. Found only in quiet waters. Leaves and flowers floating. Flowers smaller than the preceding, scarcely an inch across. Stigma dark red. July. (*Nuphar Kalmianum* R. Br.)

SARRACENIACEÆ.

SARRACENIA L.

S. purpurea L. PITCHER PLANT. SIDE-SADDLE FLOWER. HUNTSMAN'S CUP. Common in suitable places from Broome county eastward. Cinnamon Lake, Steuben county; rare, *Lucy*. Barton, *Fenno*. "The Vlai," near Oneonta; Mud and Goodrich Lakes at Davenport, *Hoy*. Near Oxford, *Coville*. Beebe's swamp; Cranberry Marsh, *Graves*. Pond Brook and Chenango Bridge, *Clute*. Hoboken Pond, Pittsfield, Otsego county, *Ellsworth*. Apalachin and Barton, *Fenno*. Usually found in sphagnum swamps or the boggy shores of our small lakes and ponds. Leaves, round, hollow, containing water and drowned insects. Flowers, dull purple, nodding; stigma umbrella-shaped; petals fiddle-shaped. There is considerable difference in the color of the leaves, which vary from deep green with light purple veining, to light greenish yellow with deep purple markings. May. June.

PAPAVERACEÆ.

SANGUINARIA L.

S. Canadensis L. BLOODROOT. Common in low meadows, on river banks and in thickets. Not frequent at Sidney, *Hoy*. Does not appear to grow in open woods within our limits. One of our earliest wild flowers; easily cultivated. Root thick, filled with crimson juice; flowers large, white; stamens yellow. Apr. May.

CHELIDONIUM L.

C. majus L. CELANDINE. SWALLOW-WORT. Not uncommon. Found along roadsides and about buildings. Has become naturalized in many places. Not reported from Tioga county. The whole plant looks not unlike a buttercup, and might be mistaken for one. Juice yellow.

———The various cultivated poppies (*Papaver*) belong here They occasionally persist about old dwellings for a few years. but may hardly be said to have become naturalized.

FUMARIACEÆ.

ADLUMIA Raf.

A. fungosa (Ait.) Greene. CLIMBING FUMITORY. MOUNTAIN FRINGE. ALLEGHANY VINE. Frequent in woods, Susquehanna county, *Graves*. Ridgebury, Bradford county, Pa.; very scarce, *Lucy*. Apparently restricted to the more mountainous parts of our range; absent elsewhere, except in cultivation. July. (*A. cirrhosa* Raf.)

BICUCULLA Adans.

B. Cucullaria (L.) Millsp. DUTCHMAN'S BREECHES. Common in low, rich woods and thickets, especially on river banks. Infrequent, *Lucy*. The curious shaped blossoms are among the most interesting of our spring flowers. April. May. (*Dicentra Cucullaria* DC.)

B. Canadensis (Goldie) Millsp. SQUIRREL CORN. Tolerably common. This species presents a gradual decline in numbers from the mountainous parts of Delaware county to the more level country west. Very common, *Hoy*. Frequent, *Graves*. Tolerably common, *Clute*. Not common, *Fenno*. Infrequent, *Lucy*. April. May. (*Dicentra Canadensis* DC.)

———The well-known cultivated Bleeding-heart (*D. spectabilis*) belongs in this genus.

CAPNOIDES Adans.

C. sempervirens (L.) Borck. PALE CORYDALIS. Found in dry, rocky soil. Not frequent at East Sidney and Davenport, *Hoy*. Frequent, *Graves*. Very rare, *Lucy*. Not noted elsewhere. (*Corydalis glauca* Pursh.)

FUMARIA L.

F. officinalis L. COMMON FUMITORY. Rare; in old gardens and waste places, *Graves*.

CRUCIFERÆ.

DENTARIA L.

D. diphylla Michx. CRINKLE ROOT. TOOTHWORT. PEPPER ROOT. Common. Found in rich, moist woods, on river banks and in ravines. Rootstocks long, creeping, toothed, peppery, edible. May.

D. laciniata Muhl. PEPPER ROOT. Common. In our territory this plant seems closely confined to the thickets along the banks of the rivers and larger streams. Apparently not so common as the preceding.

CARDAMINE L.

C. bulbosa Schreb. SPRING CRESS. Not uncommon in wet woods and the thickets along river banks. Root bearing small tubers. In our plants the flowers are white or rose-purple, without respect to the other characters which go to make up the variety *purpurea*. April. May. (*C. rhomboidea* DC.)

C. pratensis L. CUCKOO FLOWER. LADY'S SMOCK. Frequent in wet meadows, *Coville*. Tolerably common at Pond Brook and Willow Point, the only stations, *Clute*. Leaves mostly radical, the leaflets inclined to drop from the rachis easily and rooting, form new plants. Flowers conspicuous, white, or tinged with pink. May.

C. parviflora L. SMALL BITTER CRESS. Common in sandy or gravelly soil along streams. Root leaves spreading, their leaflets rounded, those of the upper leaves oblong or linear. Quite variable. May-July. (*C. hirsuta*, var. *sylvatica* Gaud.)

C. Pennsylvanica Muhl.; Willd. SPRING CRESS. Common along brooks and streams, *Lucy*.

ARABIS L.

A. laevigata (Muhl.) Poir. Infrequent. Rocky woods near Elmira, along the Chemung, *Lucy*. Unadilla Forks, *Brown*.

A. dentata T. & G. Rare. Banks of Chemung river, town of Ashland—the only station, *Lucy*.

A. Canadensis L. SICKLE-POD. Not uncommon except in the eastern part of our range. Not reported from the Chenango valley. Scarce, *Graves*. Prefers the dry, wooded slopes, where its stalks, loaded with the long pods, are very noticeable in late summer. May. June.

A. glabra (L.) Bernh. TOWER MUSTARD. Not uncommon in the eastern part of our range. Frequent, *Coville*. Not rare, *Graves*. Tolerably common, *Clute*. (*A. perfoliata* Lam.)

A. lyrata L. Rocky hillsides, *Brown*.

RORIPA Scop.

R. Nasturtium (L.) Rusby. WATER CRESS. ENGLISH WATER CRESS. Not uncommon. Streams and ponds into which it becomes introduced are rapidly choked up by its luxuriant growth. Frequently used as a salad July. (*Nasturtium officinale* R. Br.)

R. palustris (L.) Bess. MARSH CRESS. Not frequent. The typical glabrous form with oblong pods is reported from the banks of the Chemung river only, *Lucy*. (*Nasturtium palustre* DC.)

R. hispida (Desv.) Britton. MARSH CRESS. Common on river shores and in open swamps. This species is commonly recorded under *N. palustris*, but it is quite evident that it belongs rather to the variety *hispidum* of Gray's Manual, 6th edition. June. July. (*Nasturtium palustre* DC., var. *hispidum*.)

R. Armoracia (L.) A. S. Hitchcock. HORSERADISH. A common escape along streams and in wet places near dwellings. Apparently spreading. (*Nasturtium Armoracia* Fries.)

BARBAREA R. Br.

B. Barbarea (L.) Mac M. YELLOW ROCKET. COMMON WINTER CRESS. Very abundant in cultivated fields, along roadsides and other waste places. Sometimes used as a salad under the name of "poor man's cabbage." May. June. (*B. vulgaris* R. Br.)

B. præcox (J. E. Smith.) R. Br. EARLY WINTER CRESS. SCURVY GRASS. One station only—Vestal, N. Y., *Millspaugh*.

HESPERIS L.

H. matronalis L. DAME'S VIOLET. ROCKET. A rare escape that seems to be spreading. One station; scarce, *Lucy*. Not common, *Fenno*. Plentiful along the banks of the Chenango and Susquehanna near Binghamton, *Clute*. Rare near Hallstead, *Graves*. Not reported from the Chenango valley. Flowers large, white or purplish; fragrant. May. June.

ERYSIMUM L.

E. cheiranthoides L. WORM-SEED MUSTARD. TREACLE MUSTARD. Common. Said to grow in wet places, but here found along roadsides and in cultivated fields. Not reported from Susquehanna county. July.

SISYMBRIUM L.

S. officinale L. HEDGE MUSTARD. Abundant. A spreading, much-branched weed, along roadsides, about dwellings and in waste grounds. July. August.

BRASSICA L.

B. Sinapistrum Boiss. WILD MUSTARD. CHARLOCK. Found occasionally in cultivated fields, *Fenno, Lucy, Clute, Ellsworth, Coville.*

B. nigra (L.) Koch. BLACK MUSTARD. Common in cultivated ground. Flowers sweet-scented.

SINAPIS L.

S. alba L. WHITE MUSTARD. Rare. Occasionally escapes, *Barbour, Lucy.* (*Brassica alba.*)

———The cabbage (*Brassica oleracea*), turnip (*B. campestris*) and radish (*Raphnus sativus*) belong here. They sometimes appear as escapes in old gardens, but do not become naturalized.

BURSA WEBER.

B. Bursa-Pastoris (L.) Weber. SHEPHERD'S PURSE. Abundant as a weed in all cultivated grounds and waste places. Occasionally used for greens in early spring, *Graves.* Blooms throughout the spring and summer. (*Capsella Bursa-pastoris* Moench.)

LEPIDIUM L.

L. Virginicum L. PEPPERGRASS. PENNYWORT. Abundant along roadsides and in waste places.

L. campestre (L.) R. Br. YELLOW SEED. COW CRESS. Rare. Valley of the Chenango, *Clute.* Well established along Seeley creek near Elmira, *Lucy.* Not noted elsewhere. June.

CAPPARIDACEÆ.

POLANISIA Raf.

P. gravolens Raf. River bank near Lanesboro; not rare, *Graves*. Shores of Chemung river; frequent, *Lucy*. Not found elsewhere.

CISTACEÆ.

HELIANTHEMUM Pers.

H. Canadense (L.) Michx. Rock Rose. Frost Weed. In gravelly soil; river bank at Apalachin; not common, *Fenno*. Plant with the aspect of a St. John's-wort. Flowers of two sizes; petals of the large, early ones very fugacious. Later in the year minute flowers are produced in the axils of the leaves.

LECHEA L.

L. villosa Ell. Pinweed. Rare. Near a roadside in an open glade, Chemung county, *Lucy*. Ely Hill, *Millspaugh*. (*L. major* Michx.)

L. intermedia Leggett; Britton. Rare. Near Elmira, *Lucy*.

VIOLACEÆ.

VIOLA L.

V. pedata L. Bird-foot Violet. Rare. Ross Park; east decline House's Hill, *Millspaugh*.

V. palmata L. Hand-leaved Violet. Blue Violet. Not very common. Usually found in upland woods. Leaves variously cut and divided. Flower large, deep blue. May. June.

V. palmata cordata Walt. Rare. Glenwood ravine, Binghamton; Spanish Hill, Waverly, *Millspaugh*.

V. obliqua Hill. Common Blue Violet. Abundant in wet meadows, and along streams. Our commonest Violet. Produces numerous cleistogamous flowers in summer. May. June. Blooms freely again in October. (*V. palmata* L., var. *cucullata* Gray.)

V. sagittata Ait. Arrow-leaved Violet. Unequally distributed within our range. Frequent in sandy soil, *Hoy*. Dry roadsides;

scarce, *Graves*. Common on dry hills and along roadsides, *Clute*. Frequent west of Barton, *Fenno*. Common near New Berlin, *Ellsworth*. Common in sandy meadows, Sayre, *Barbour*. Cayuta creek, *Millspaugh*. Sparingly on dry hillsides at Oxford, *Coville*. Not reported from Tioga and Chemung counties. Flowers as large as those of the two preceding. Cleistogamous flowers are produced in summer. May. June.

V. Selkirkii Pursh. SELKIRK'S VIOLET. GREAT SPURRED VIOLET. Very rare. Hill at Vestal, *Millspaugh*. Along a brook near Oxford, *Coville*.

V. blanda Willd. SWEET WHITE VIOLET. Common in swamps, along streams and in wet, open woods. Flowers very fragrant. The plant spreads by leafy runners which usually bear cleistogamous flowers in the axils of the leaves. April. May.

V. blanda amœna (Le Conte) B. S. P. Reported occasionally. Lowman's swamp, town of Chemung, *Lucy*. Ravine, Choconut Centre, *Clute*. This species is often referred to *V. blanda*. A closer study of these two will doubtless reveal the sub-species at other points within our limits. (*V. blanda*, var. *palustriformis* A. Gray.)

V. rotundifolia Michx. ROUND-LEAVED YELLOW VIOLET. Tolerably common. Found in cold, wet soil, especially in ravines. Grows also in open woods. Infrequent, *Lucy*. Not reported from Susquehanna county. One of our earliest and handsomest species. Spreads by runners. April.

V. pubescens Ait. DOWNY YELLOW VIOLET. STEMMED YELLOW VIOLET. Very common in damp woodlands. Produces cleistogamous flowers in summer. May.

V. scabriuscula (T. & G.) Schwein. ROUGH-LEAVED YELLOW VIOLET. Reported rare. South Mountain, *Millspaugh*. Greatsinger's corners, Chemung county; plentiful, *Lucy*. Occasionally found, *Clute*. This has doubtless been overlooked at other points in our range, and should be searched for. Plants smaller, greener, and slightly if at all pubescent. (*V. pubescens*, var. *scabriuscula* T. & G.)

V. pubescens eriocarpa (Scher.) Nutt. Frequent, *Coville*. Near Binghamton, *Millspaugh*.

V. Canadensis L. Canada Violet. Common in moist woodlands, especially in ravines. Blooms all summer and early autumn. Flowers white or purplish, the back or outside of petals usually deep purple; faintly fragrant. Grows well in cultivation. Our tallest violet

V. striata Ait. Pale Violet. Striped Violet. Not very common. Most frequent on the river flats, also in open woods. Not reported from Delaware county. Common in moist meadows near Oxford, *Coville*. Occasionally blooms in November, *Graves*. May. June.

V. rostrata Pursh. Long-spurred Violet. Unequally distributed. Common, *Hey*. Frequent, *Graves*. Plentiful, *Clute*. Not common, *Fenno*. Rare, *Lucy*. A beautiful species in moist, rich woods. Characterized by its very long spur. Blooms all through the summer. Excellent for cultivation in shade.

V. Labradorica Schrank. Dog Violet. Common. Found in open woods and thickets, wet or dry. Next to *V. obliqua* this is our most abundant Violet. Most widely distributed of the genus. Occasionally blooms in autumn. April. May. (*V. canina* L., var. *Muhlenbergii* Gray.)

V. tricolor L. Pansy. Heartsease. A rare escape, recorded from Chenango, Susquehanna, Broome and Chemung counties. It is doubtful if this species persists long with us in the wild state.

CARYOPHYLLACEÆ.

DIANTHUS L.

D. barbatus L. Sweet William. Cultivated, but occasionally escapes. Rare; along roadsides.

SAPONARIA L.

S. officinalis L. Bouncing Bet. Soapwort. Very common in roadsides and waste places, especially along railroads and banks of streams. Flowers nearly all summer.

SILENE L.

S. stellata (L.) Ait. Starry Campion. Not common except in the central part of our territory. Prefers rather dry, shaded banks, especially open, upland woods. Infrequent, *Lucy*. Not common, *Barbour*. Common, *Fenno*. Abundant, *Clute*. Scarce, *Graves*. Not reported from Delaware county. Flowers white, the petals fringed. A beautiful plant. July.

S. vulgaris (Moench) Garcke. BLADDER CAMPION. Rare. Sometimes escapes from cultivation. Scarce, *Graves*. Rare, *Brown*. Occasional, *Clute*. It is said that the young shoots and leaves may be used as a substitute for asparagus. (*S. Cucubalus* Wibel.)

S. antirrhina L. SLEEPY CATCHFLY. Infrequent. Along the railroad, Ashland, Chemung county, *Lucy*.

S. antirrhina divaricata Robinson. Rare. Plants found on a dry wooded slope near Binghamton in 1896 were identified as this sub-species at the Harvard Herbarium, *Clute*.

S. noctiflora L. NIGHT-FLOWERING CATCHFLY. Somewhat rare. Reported from their localities by *Coville*, *Lucy*, *Graves* and *Clute*.

LYCHNIS L.

L. Chalcedonica L. MALTESE CROSS. SCARLETLYCHNIS. Occasional. *Lucy*.

L. dioica L. RED LYCHNIS. Scarce; in dry soil, *Graves*. (*L. diurna* Sibth.)

AGROSTEMMA L.

A. Githago L. CORN COCKLE. WOOLLY PINK. ROSE CAMPION. Not common. In grain fields and occasionally in waste places. Not reported from Delaware county. (*Lychnis Githago* Lam.)

ARENARIA L.

A. serpyllifolia L. THYME-LEAVED SANDWORT. Unequally distributed. Reported common by *Hoy*, *Clute* and *Lucy*; scarce, *Coville*. Not noted elsewhere. An interesting little plant inhabiting dry, sandy soils. Stems, much branched, six inches high; leaves, minute, ovate, ciliate; flowers, numerous, white. June. July.

A. lateriflora L. LATERAL-FLOWERED SANDWORT. Common at Sayre, *Barbour*. Tolerably common at Kirkwood and Willow Point, *Clute*. Scarce, *Graves*. Not noted elsewhere. May. June.

ALSINE L.

A. media L. COMMON CHICKWEED. An abundant weed in cultivated and waste grounds. Flowers from earliest spring to autumn. Closes its flowers at the approach of storms, *Graves*. The earliest flowers are fertilized in the bud. (*Stellaria media* Smith.)

A. longifolia (Muhl.) Britton. LONG-LEAVED STITCHWORT. Common in moist, grassy places. June. (*Stellaria longifolia* Muhl.)

A. borealis (Bigel.) Britton. NORTHERN STITCHWORT. Rare. Wet place at base of wooded hill near North Chemung, *Lucy*. Rather rare. In wet places, *Graves*. Oxford, *Coville*. Not noted elsewhere. (*Stellaria borealis* Bigel.)

A. graminea (Muhl.) Britton. GRASS CHICKWEED. Rare. Wet ditch along a roadside; flowers large and conspicuous, *Graves*, (*Stellaria graminea* L.)

CERASTIUM L.

C. vulgatum L. MOUSE-EAR CHICKWEED. A common and well-known weed in cultivated grounds and waste places. Flowering nearly all summer.

C. longipedunculatum Muhl. CHICKWEED. Base of Sullivan Hill and along the Chemung river. In specimens here, the stem is strict and leafy, *Lucy*. (*C. nutans* Raf.)

SPERGULA L.

S. arvensis L. CORN SPURRY. A common weed in gardens. Scarce, *Lucy*. Stem, round, with swollen joints, branched; leaves, thread-like in whorls.

PORTULACACEÆ.

PORTULACA L.

P. oleracea L. PURSLANE. PUSLEY. Abundant as a weed in cultivated grounds, also in all waste places. Very tenacious of life, after flowering it will ripen its seeds though severed from its roots. Sometimes used for a pot-herb. July–Sept.

CLAYTONIA L.

C. Virginica L. SPRING BEAUTY. Common in wet woods, thickets and especially the low meadows along streams. The stems arise from a round, flat tuber rather deep in the earth, usually several stems from each tuber. May.

C. Caroliniana Michx. SPRING BEAUTY. Apparently less common than the preceding. Not reported from Susquehanna and Broome counties. Seems to prefer greater elevations, and is often found in open, wet woods on hills. Distinguished from *Virginica* by its shorter, broader leaves, and smaller flowers. May.

HYPERICACEÆ.

HYPERICUM L.

H. Ascyron L. GREAT ST. JOHN'S WORT. Very common along the rivers. Occasionally found in swamps. Scarce, *Coville*. The largest member of the genus, often growing to a height of six feet, with flowers two inches across. July.

H. ellipticum Hook. ELLIPTIC-LEAVED ST. JOHN'S WORT. Tolerably common, especially in gravelly soil along streams. Petals often but four. Much resembles *H. perforatum*. July. Aug.

H. perforatum L. COMMON ST. JOHN'S WORT. Abundantly naturalized in fields, pastures, roadsides and waste places. Spreads by runners. June–Sept.

H. maculatum Walt. CORYMBED ST. JOHN'S WORT. SPOTTED ST. JOHN'S WORT. BLACK-DOTTED ST. JOHN'S WORT. Common in shade. Roadsides, thickets and open woods.

H. mutilum L. DWARF ST. JOHN'S WORT. Common, especially in damp, sterile soil; fields and roadsides. Stems seldom more than six inches high, with numerous small, copper-colored flowers. July.

H. Canadense L. CANADIAN ST. JOHN'S WORT. Chenango valley and near Waverly, *Millspaugh*.

H. Virginicum L. MARSH ST. JOHN'S WORT. Common in swamps and along streams. Not reported from the valley of the Chemung. Flowers flesh-colored in the axils of the leaves, seldom opening in bright sunlight. Spreads by underground runners. July. Aug. (*Elodes campanulata* Pursh.)

MALVACEÆ.

MALVA L.

M. rotundifolia L. Low Mallow. Round-leaved Mallow. Cheeses. A common weed in fields and about buildings. The heads of fruit contain considerable mucilage, and when green are frequently eaten by children.

M. sylvestris L. High Mallow. A rare escape. Reported from Otsego, Chenango, Broome and Steuben counties.

M. moschata L. Musk Mallow. Common along roadsides and in meadows. Flowers large, as frequently white as rose-colored. The whole plant has a musky odor. July. Sept.

ABUTILON Gærtn.

A. Abutilon (L.) Rusby. Indian Mallow. Velvet Leaf. An escape that is fast becoming naturalized in cultivated grounds. Rare in Chemung county, common farther north, *Lucy*. Common, *Clute*. Not common, *Coville*. Not reported elsewhere. (*A. Avicennæ* Gærtn.)

HIBISCUS L.

H. Trionum L. Bladder Ketmia. Very rare. Wellsburg; Ashland, *Lucy*.

TILIACEÆ.

TILIA L.

T. Americana L. Basswood. Linden. Linn. Whitewood. Common throughout in rich soil. A large tree with light, soft wood. Bees obtain much honey from the blossoms.

LINACEÆ.

GERANIACEÆ.

GERANIUM L.

G. maculatum L. CRANESBILL. WILD GERANIUM. Common. Found in woods, thickets and moist meadows. This and the following species are very easily cultivated. May.

G. Robertianum L. HERB-ROBERT. Not common except in damp, rocky woods, and not always found in such situations within our limits. Infrequent in the Chemung valley, *Lucy.* Common in the Chenango valley north of Broome county, *Coville; Brown.* Also in the Susquehanna valley east of Broome county. Not found in the valley from Binghamton west. Plentiful just south of our limits. June–Sept.

G. Carolinianum L. Rare. Ashland and Elmira, *Lucy.* South Mountain, *Millspaugh.*

OXALIDACEÆ.

OXALIS L.

O. Acetosella L. COMMON WOOD SORREL. Common in dark, cold woods, especially under hemlocks. Flower rather large, white veined with purple. June–July.

O. violacea L. VIOLET WOOD SORREL. Rare. Near Waverly, *Millspaugh.* River bank near Apalachin; common, *Fenno.* Plentiful on a rocky hillside near Binghamton, *Clute.* Thrives in bright sunshine. Bulb, scaly; flowers, several in an umbel, pink, not "violet-colored" with us. The plant spreads by means of underground runners which produce bulbs at the ends. May.

O. stricta L. SHEEP SORREL. YELLOW WOOD SORREL. SOUR GRASS. Abundant in cultivated and waste grounds. Flowers, small, yellow. The leaves of this and the two preceding species have a pleasant acid taste, and are often eaten by children. (*O. corniculata,* var. *stricta* Sav.)

BALSAMINACEÆ.

IMPATIENS L.

I. aurea Muhl. PALE TOUCH-ME-NOT or JEWEL WEED. SNAP WEED. SNAPDRAGON. Common in damp, shady situations. Found along river banks and in wet, open woods, often in company with the following species. Flowers large, pale yellow. July. Aug. (*I. pallida* Nutt.)

I. biflora Walt. SPOTTED TOUCH-ME-NOT or JEWEL WEED. SNAPDRAGON. SILVER WEED. Very common in wet grounds in the shade or sun. Somewhat more common than the preceding. Flowers similar in shape, smaller, orange, spotted with reddish brown, especially on the lower lip. Spotless flowers sometimes occur. A form of this blossom with blood-red or pink lower lip is not infrequent. Called Silver Weed from the silvery appearance of the leaves when plunged under water. The capsules of this and the preceding species, when mature, burst at the slightest touch, scattering the seeds. Cleistogamous flowers are usually produced before the showy ones. July. Aug. (*I. fulva* Nutt.)

RUTACEÆ.

XANTHOXYLUM L.

X. Americanum Mill. PRICKLY ASH. TOOTHACHE TREE. Somewhat rare. Prefers moist soil. Town of Ashland, Chemung county, *Lucy*. West of Apalachin, *Fenno*. Susquehanna, *Graves*. Norwich, *Coville*. Unadilla Forks, *Brown*.

PTELEA L.

P. trifoliata L. HOP-TREE. SHRUBBY TREFOIL. Very rare. A single specimen is growing (1896) along Riverside Drive in the city of Binghamton, where it is apparently indigenous, *Clute*.

The Chinese Tree of Heaven (*Ailanthus*) belongs here. It is occasionally cultivated for shade and is said to have a tendency to become naturalized.

ILICINEÆ.

ILEX L.

I. verticillata (L.) A. Gray. WINTERBERRY. BLACK ALDER. DECIDUOUS HOLLY. Common in open swamps and low grounds. Flowers small, axillary. May. June. Fruit ripe in September. After the leaves have fallen the berries are very noticeable and render this shrub one of the most conspicuous objects in our winter landscape. A form with orange-colored berries is sometimes found. The berries are seldom if ever eaten by birds.

I. lævigata (Pursh.) A. Gray. SMOOTH WINTERBERRY. Not rare in swamps near Susquehanna, *Graves*.

ILICIOIDES DUMONT.

I. mucronata (L.) Britton. MOUNTAIN HOLLY. Common only in the more elevated parts of our range, in cold, moist soil. Frequent, *Graves*. The Vlai, Goodrich and Sexsmith lakes, common, *Hoy*. Rare, *Clute*. Not common; Mutton Hill Pond, *Fenno*. Common about sphagnum bogs, *Coville*. Not reported elsewhere. Has much the aspect of a small Shad-bush (*Amelanchier*) with usually solitary red berries on long pedicels. May. (*Nemopanthes fascicularis* Raf.)

CELASTRACEÆ.

CELASTRUS L.

C. scandens L. BITTERSWEET. WAXWORK. STAFF-TREE. Common about old fences, and along streams. A well-known half-shrubby climber, bearing orange-colored pods which open in autumn displaying the scarlet covering of the seeds within. An attractive plant, easily cultivated. The fruit is valued for winter decorations.

EUONYMUS L.

E. Americanus L. STRAWBERRY BUSH. Very rare. A single shrub found in Susquehanna; perhaps cultivated, *Graves*.

E. obovatus Nutt. STRAWBERRY BUSH. Rare. Two miles south of Apalachin, *Fenno*. Trailing with rooting branches. (*E. Americanus*, var. *obovatus* T. & G.)

RHAMNACEÆ.

RHAMNUS L.

R. alnifolia L'Her. BUCKTHORN. Rare. Baldwin creek, *Lucy*. Suburbs of Binghamton, *Clute*. Goodrich Lake, *Hoy*. A low, unarmed shrub, with ill-smelling foliage.

R. cathartica L. COMMON BUCKTHORN. Noted from but two localities. Apalachin, common in hedges, *Fenno*. Hill south of Binghamton; rare, *Clute*.

CEANOTHUS L.

C. Americanus L. NEW JERSEY TEA. RED-ROOT. Abundant on all our hills, either in sun or shade. Apparently no hillside is too dry for it. Root perennial, very large in proportion to the rest of the plant, red; sometimes used in coloring. Flowers in clusters, small, white, numerous, making the plant very conspicuous when in bloom. Dies down nearly to the ground in winter. Few plants are better than this for a low hedge that will take care of itself and look well at all seasons. The principal plant on many of our barren hillsides. Leaves said to have been used for tea. At maturity the capsules burst, scattering the seeds with some force. June.

VITACEÆ.

VITIS L.

V. Labrusca L. NORTHERN FOX GRAPE. Very rare. Valley of the Chenango, *Millspaugh*. The original of most of our cultivated grapes.

V. æstivalis Michx. SUMMER GRAPE. Common. Along river banks and fence rows. Rare in Chenango county, *Coville*.

V. cordifolia Michx. FROST GRAPE. CHICKEN GRAPE. Abundant, especially along the banks of streams, where it fruits heavily each year. Berry small, inedible until frosted. Flowers with a strong spicy fragrance.

V. vulpina L. FROST GRAPE. Common in the Chemung valley, *Lucy*. Sidney; Goodrich Lake, *Hoy*. Not reported elsewhere. It is evident that our grapes need further study. (*V. riparia* Michx.)

PARTHENOCISSUS Planch.

P. quinquefolia (L.) Planch. WOODBINE. VIRGINIA CREEPER. Abundant in woods, thickets and along fence rows. Common in cultivation. Often mistaken for poison ivy Fruit purplish black. Leaves turning crimson in autumn. (*Ampelopsis quinquefolia* Michx.)

HIPPOCASTANACEÆ.

ÆSCULUS L.

Æ. Hippocastanum L. HORSE CHESTNUT. Common in cultivation for shade, and occasionally escapes.

ACERACEÆ.

ACER L.

A. Pennsylvanicum L. STRIPED MAPLE. MOOSEWOOD. WHISTLEWOOD. STRIPED DOGWOOD. Common. Most plentiful in ravines and along cliffs. Readily distinguished by its smooth, light-green bark, heavily striped with ashy-green. The bark of the trunk does not become rough except in the oldest individuals. With us, it is seldom more than a shrub; but there are a few trees within our limits with trunks eight inches in diameter.

A. spicatum Lam. MOUNTAIN MAPLEBUSH. SPIKED MAPLE. GOOSEFOOT MAPLE. Abundant; forming thickets along mountain streams, in glens and ravines.

A. Saccharum Marsh. SUGAR MAPLE. ROCK MAPLE. HARD MAPLE. Common in woods and the open fields. The species most commonly used for a shade tree. From the sap of this and the following is produced most of our maple sugar. (*A. saccharinum* Wang.)

A. nigrum Michx f. BLACK MAPLE. Not so common as the preceding with which it is very often confounded. May be distinguished by its darker bark. Blooms earlier. (*A. saccharinum* var. *nigrum* T. & G.)

A. saccharinum L. SILVER MAPLE. WHITE MAPLE. RIVER MAPLE. Very common. Almost restricted to the banks of streams.

Leaves more deeply lobed than in any of our other common species. Sap thin, watery, sour. Occasionally planted for shade. (*A. dasycarpum* Ehrh.)

A. rubrum L. RED MAPLE. SOFT MAPLE. SWAMP MAPLE. Common. Does not grow exclusively in moist places, but may be found on hill-tops. Noticeable in spring for its red blossoms, and again in autumn for its leaves which then turn scarlet. The sap yields sugar. Often planted for shade.

A. Negundo L. ASH-LEAVED MAPLE. BOX ELDER. Rare. Frequently cultivated about Binghamton, where it is commonly found as an escape, *Clute*. Often cultivated for shade, *Fenno*. Becoming common in cultivation, *Brown*. (*Negundo aceroides* Moench.)

STAPHYLEACEÆ.

STAPHYLEA L.

S. trifolia L. BLADDERNUT. Chenango river, Binghamton, *Millspaugh*. Occasional at Apalachin; very abundant at Barton, *Fenno*. Not noted elsewhere.

ANACARDIACEÆ.

RHUS L.

R. hirta (L.) Sudw. STAG-HORN SUMAC. Very common, especially on dry, rocky hillsides. Our tallest species often reaching the size of a small tree. Young branches velvety-hairy. Leaves turning crimson in autumn. Fruit a small, berry-like drupe with crimson down, collected into thick, close panicles known as "sumac bobs," very acid, edible. The wood is beautifully grained with shades of brown, green and yellow, and the bark contains much tannin. (*R. typhina* L.)

R. glabra L. SMOOTH SUMAC. Less common than the preceding. Not reported from the upper Chenango valley. Rather scarce, *Hoy*. Found in the same situations as *R. hirta*, often intermixed with it. Shrub lower, smooth, otherwise much like that species. The fruit is sometimes eaten.

R. copallina L. MOUNTAIN SUMAC. DWARF SUMAC. Rare. Ross Park, *Millspaugh*.

R. Vernix L. POISON SUMAC. Unequally distributed. Scarce, *Lucy*. Butler's Lake—the only station, *Graves*. Brisbin swamp, *Coville*. Not reported from the northern part of the Susquehanna and Chenango valleys. Common from Binghamton west. Found in wet woods, borders of swamps and occasionally in dry ground. A shrub or small tree, of elegant aspect, but very poisonous to the touch, and even tainting the air for some distance with its poison. The poisonous effects are said to be due to a volatile acid. Rochelle salts internally and an application of "soft" soap to the poisoned parts will quickly effect a cure. Herbivorous animals are said to eat the leaves of this species without injury, and many persons are not affected by the poison. Distinguished from the other species by its entire leaflets. Drupe, gray. Also known as Poison Oak, Poison Elder and Poison Dogwood. (*R. venenata* DC.)

R. radicans L. POISON IVY. THREE-LEAVED MERCURY. POISON OAK. Common in both wet and dry soils. Where this plant finds a support it often becomes a vine several inches in diameter, climbing to the tops of trees and anchored to the trunk by numerous rootlets. Unsupported it takes on a shrubby form and is then most frequently called Poison Oak. Berries gray, in loose racemes. Leaflets three. The juice forms an indelible ink. Poisonous, but less so than the preceding. (*R. toxicodendron* L.)

R. aromatica Ait. AROMATIC SUMAC. SWEET SUMAC. Choconut Creek, *Millspaugh*. Not rare, *Graves*. Frequent, *Lucy*. Not reported elsewhere. A small shrub on dry banks, with three-parted, aromatic leaves and flowers in close aments appearing in early spring. Not poisonous. (*R. Canadensis* Marsh.)

POLYGALACEÆ.

POLYGALA L.

P. paucifolia Willd. FLOWERING WINTERGREEN. INDIAN PINK. BABY FOOT. FRINGED POLYGALA. Common in thickets and open woods. Less common in the northeastern part of our range. An elegant little plant with the aspect of a Wintergreen. Flowers two or more, rather large, pink-purple. Cleistogamous flowers are borne at the base of the plant. A white-flowered variety is occasionally noticed. May.

P. Senega L. SENECA SNAKEROOT. Somewhat rare. Found on dry, scrubby hilltops. Infrequent, *Lucy*. Near Apalachin, *Millspaugh*. Hills about Binghamton; not uncommon, *Clute*. Rare, *Graves*. Not reported northeast of Broome county. Flowers in terminal spikes, white. June.

P. viridescens L. RED MILKWORT. Not common, *Graves*. Hills south of Apalachin, *Clute*. Flowers red, in dense spikes. The root is said to have an odor like that of the aromatic Wintergreen. (*P. sanguinea* L.)

P. verticillata L. WHORLED POLYGALA. Common from Susquehanna county west. Not reported from the northern part of the Chenango and Susquehanna valleys. Found in dry situations especially hillsides in grass. Flowers minute, purplish-white in spikes. Easily overlooked. Summer.

LEGUMINOSÆ.

BAPTISIA VENT.

B. tinctoria (L.) R. Br. FALSE INDIGO. Very rare. Gulf Summit, *Millspaugh*.

LUPINUS L.

L. perennis L. LUPINE. SUN-DIAL. Abundant in Broome and Tioga counties, *Clute*, *Fenno*. Infrequent, *Lucy*. Not rare, *Graves*. Not reported elsewhere within our limits. Found along streams, on hillsides, in thickets and borders of the woods. Very noticeable when in bloom. Flowers large, blue and purple, in long terminal spikes. A white-flowered form is occasionally reported. Flowers sometimes slightly fragrant. June.

TRIFOLIUM L.

T. arvense L. RABBIT'S-FOOT or STONE CLOVER. Sullivan Hill; Cobble Hill, Chemung county; plentiful at stations, *Lucy*. Barton; Apalachin, *Fenno*. Binghamton, *Ellsworth*.

T. pratense L. RED CLOVER. Abundant, usually in cultivation.

T. reflexum L. BUFFALO CLOVER. Rare. South Mountain, *Millspaugh*.

T. repens L. WHITE CLOVER. SHAMROCK. Abundant everywhere. Believed to be one of the first plants to appear in burned areas.

T. hybridum L. ALSIKE CLOVER. HONEY CLOVER. Becoming common. Flowers rose-tinted; stems not rooting at the joints, otherwise much like the preceding.

T. agrarium L. YELLOW CLOVER. HOP CLOVER. Abundantly naturalized in dry fields, along roadsides, etc. A pretty clover, with bright yellow flower-heads, somewhat resembling hops.

T. procumbens L. DWARF HOP CLOVER. Very common in Tioga and Broome counties, for the most part restricted to the railways, growing between the ties, *Fenno*, *Clute*. Roadsides, common, *Brown*. Not reported elsewhere. Plant procumbent, spreading, with flower-heads one-fourth the size of the preceding.

T. incarnatum L. CRIMSON CLOVER. Rare. Is occasionally planted for forage and in time will doubtless become common. Apalachin, *Fenno*. Binghamton; Tunkhannock, *Clute*.

MELILOTUS Juss.

M. officinalis (L.) Lam. YELLOW MELILOT. YELLOW SWEET CLOVER. Rare in the eastern part of our range; more common west. In waste grounds, wet or dry.

M. alba Lam. WHITE SWEET CLOVER or MELILOT. Very common throughout, growing in the same places as the preceding and especially along railways and river banks. The flowers and leaves of both emit a strong, sweetish odor when drying.

MEDICAGO L.

M. sativa L. ALFALFA. LUCERNE. Found occasionally in Broome county, *Clute*. Flowers bluish-purple, racemed. Pods spirally coiled.

M. lupulina L. BLACK MEDICK. NONESUCH. Infrequent; banks of the Chemung near Elmira, *Lucy*. Becoming a weed in lawns, *Coville*. Flowers yellow.

ROBINIA L.

R. pseudacacia L. COMMON LOCUST. FALSE ACACIA. Plentiful, especially near streams. Flowers large, white, sweet-scented, in racemes. The wood is famed for its durability in exposed situations; fence posts of this wood have been known to last for more than seventy years.

ASTRAGALUS L.

A. Carolinianus L. MILK VETCH. Rare. Canal bank north of Port Crane, *Millspaugh*. Hemlock Lake, Livingston county, N. Y., *Lucy*.

CORONILLA L.

C. varia L. Near Pine Valley, four miles above Horseheads, in a moist, grassy place; appears to be spreading, *Lucy*.

MEIBOMIA ADANS.

M. nudiflora (L.) Kuntze. TICK TREFOIL. Common in dry woods. Flowers, medium sized, on a long, leafless scape from the root. (*Desmodium nudiflorum* DC.)

M. grandiflora (Walt.) Kuntze. TICK TREFOIL. Common in woods. Leaves all clustered at the top of the stem, from whence the flower stalk rises. (*Desmodium acuminatum* DC.)

M. pauciflora (Nutt.) Kuntze. Waverly, N. Y., 1862, *Graves*. Near Elmira, *Lucy*. (*Desmodium pauciflorum* DC.)

M. rotundifolia (Michx.) Kuntze. ROUND-LEAVED TICK TREFOIL. Rather rare. Sullivan Hill, *Lucy*. Hill northeast of Binghamton, *Millspaugh*. Hill above Oakland, *Graves*. Not noted elsewhere. (*Desmodium rotundifolium* DC.)

M. Dillenii (Darl.) Kuntze. Occasional in dry fields, *Graves*. Not uncommon in copses, *Clute*. Sullivan Hill; infrequent, *Lucy*. (*Desmodium Dillenii* Darl.)

M. paniculata (L.) Kuntze. Common in copses. July. (*Desmodium paniculatum* DC.)

M. Canadensis (L.) Kuntze. BUSH TREFOIL. Abundant along our river shores, where the tall stems often form dense thickets. July. (*Desmodium Canadense* DC.)

M. Marylandica (L.) Kuntze. Rare. Sullivan Hill, *Lucy*. Near Waverly, N. Y., *Graves*. (*Desmodium Marlandicum* F. Boott.)

The range of the various species of *Meibomia* within our limits are not yet clearly defined. This group needs further study.

LESPEDEZA Michx.

L. violacea (L.) Pers. Frequent in dry copses, *Fenno*. Not uncommon, *Clute*. Chenango county, *Coville*.

L. hirta (L.) Ell. BUSH CLOVER. Common in dry thickets and along roadsides. Scarce, *Lucy*. Flowers cream-colored with a purple spot on the standard. Aug. (*L. polystachya* Michx.)

L. capitata Michx. BUSH CLOVER. Elmira, Big Flats and Corning; not rare at stations, *Lucy*. Susquehanna, *Graves*.

VICIA L.

V. Cracca L. TUFTED VETCH. Not common, *Coville*. Frequent, *Graves*. Abundant, *Clute*. Not reported elsewhere. Found along roadsides and in dry fields. May.

V. Caroliniana Walt. VETCH or TARE. Our most abundant species. Found in fields and thickets. May.

V. Americana Muhl.; Willd. Reported only from Tioga county west. Apalachin, *Fenno*. East Waverly, *Millspaugh*. Valley of the Chemung, *Lucy*.

LATHYRUS L.

L. ochroleucus Hook. VETCHLING. EVERLASTING PEA. CREAM-COLORED LATHYRUS. Usually found in woods. Scarce, *Graves*. Plentiful, *Clute*. Common, *Lucy*. Elsewhere not reported. Vine climbing over other herbage; flowers rather large.

APIOS Moench.

A. Apios (L.) MacM. GROUND-NUT. WILD BEAN. TUBEROUS WISTARIA. Not uncommon throughout. Found on river banks and in other damp situations, climbing over the surrounding plants. Flowers in dense racemes, pink-brown in color, very fragrant. Edible tubers are borne on underground shoots. Although provided with excellent means for cross-fertilization, this plant seldom sets seed, and spreads mainly by its subterranean runners. Frequently cultivated for shade. Aug. (*A. tuberosa* Moench.)

FALCATA Gmel.

F. comosa (L.) Kuntze. HOG PEANUT. PEA VINE. Very common in woods, especially in damp soil. A low vine, with racemes of small, pink or white flowers, which seldom set seed. At the base of the plant are smaller apetalous flowers, which are commonly fruitful, producing a fleshy pod containing a single seed. (*Amphicarpæa monoica* Nutt.)

CASSIA L.

C. Marylandica L. WILD SENNA. Rare. Reported only from the Chenango valley. Along a roadside, South Oxford, *Coville*. On a low island in the Chenango at Port Dickinson, where it is not uncommon, *Clute*. One specimen found at Wellsburg, Chemung county, in 1874, *Lucy*. A beautiful, half shrubby, locust-like plant, with racemes of nearly regular, bright yellow flowers. Leaf stem with a club-shaped gland at base. July.

GLEDITSCHIA L.

G. triacanthos L. HONEY LOCUST. THREE-THORNED ACACIA. Common in Susquehanna and Broome counties, where it is frequently used for hedges. Planted for shade, it forms a large tree with spreading, open top. Pods very long and broad.

ROSACEÆ.

PRUNUS L.

P. Americana Marsh. WILD YELLOW or RED PLUM. Not uncommon. Prefers damp situations, and is usually found along streams or in moist woodlands. Sometimes attains the size of a small tree. Fruit sweet, edible.

P. Pennsylvanica L. f. WILD RED CHERRY. PIN CHERRY. BIRD CHERRY. FIRE CHERRY. Common. A small tree, of quick growth, found in thickets and along fences. Fruit very small, sour, light red. Called "Fire Cherry" from the rapidity with which it appears in burned tracts. What is apparently a very dwarf form of this is reported from the vicinity of Susquehanna by Mr. Graves. It is shrubby, two feet or more high, and well fruited. Fruit like the species.

P. Virginiana L. CHOKE CHERRY. Very common in fence rows, along roadsides and the banks of streams. A tall shrub. Flowers in racemes, followed by an abundance of dark crimson fruit, very astringent and scarcely edible.

P. serotina Ehrh. BLACK CHERRY. Rather common in woodlands. A large tree, with strong, red wood. Fruit in racemes, black when ripe, sweet, edible.

——The peach, plum, apricot and cherry of our gardens belong in this genus. They occasionally escape from cultivation, especially the peach and sour cherry, and are found along roadsides and in waste grounds.

SPIRÆA L.

S. salicifolia L. MEADOW SWEET. Common, especially in low grounds. A well-known inhabitant of meadows and swamps. July.

S. tomentosa L. HARD-HACK. STEEPLE-BUSH. Frequent, *Graves*. Not uncommon, *Clute*. Not recorded elsewhere. Found in low meadows and pastures. Stem and under side of leaves covered with a rusty down. Flowers rose-colored, in a dense terminal panicle. Called hard-hack by the haymakers from its hard, brittle stems. "The persistent fruit in winter furnishes food for the snowbird."—Wood's *Class Book of Botany*. Aug.

OPULASTER MEDIC.

O. opulifolius (L.) Kuntze. NINE-BARK. Common only in Tioga and Broome counties. Occasional, *Graves*. Rare, *Lucy*. Not noted elsewhere. Found on river banks. Bark deciduous, stringy; leaves slightly three-lobed; flowers white, in corymbs, succeeded by conspicuous membranaceous pods. Excellent for cultivation. June. (*Physocarpus opulifolius* Maxim.)

PORTERANTHUS BRITTON.

P. trifoliatus (L.) Britton. INDIAN PHYSIC. BOWMAN'S ROOT. Common from Broome county west. Not recorded from other parts of our range. An interesting plant in open upland woods. Flowers white; petals narrow; leaves three-foliate; stipules small, awl-shaped. A form of this with ovate or obovate incised stipules is common about Binghamton, growing with the other, *Clute*. (*Gillenia trifoliata* Mœnch.)

RUBUS L.

R. odoratus L. PURPLE-FLOWERING RASPBERRY. THIMBLE-BERRY. MULBERRY. Common in cool rocky woods, especially in ravines. Well-known for its showy blossoms. Fruit nearly flat, insipid. June.

R. Americanus (Pers.) Britton. DWARF RASPBERRY. Frequent, especially in low woods. Fruit consists of a few large, dark-red grains. "This appears to be more properly a blackberry."— Gray's *Manual*. (*R. triflorus* Richards.)

R. strigosus Michx. WILD RED RASPBERRY. Common throughout in woods, along roadsides, and especially in slashings. Well known and valued for its fruit.

R. occidentalis L. BLACK RASPBERRY. BLACK-CAP. THIMBLE-BERRY. As well distributed as the preceding, but in somewhat lesser numbers. Found in woods, thickets and fence-rows. Usually very fruitful.

R. villosus Ait. HIGH BLACKBERRY. Abundant in woods, thickets, and especially in clearings.

R. villosus frondosus Torr. SMOOTH-STEMMED BLACKBERRY. Occasional, *Graves*. Frequent, *Lucy*. The only recorded localities within our range. Plant smoother, less glandular, and with larger bracts than the species, which it otherwise closely resembles. (*R. villosus*, var. *frondosus* Torr.)

R. Canadensis L. DEWBERRY. LOW BLACKBERRY. RUNNING BLACKBERRY. Common in fields and on hillsides. Stem prostrate. Fruit consisting of numerous large, black grains, each enclosing a hard seed. Less valued than the high blackberry.

R. hispidus L. RUNNING SWAMP BLACKBERRY. Plentiful in low, open woods and swamps. Fruit of little consequence. Leaves persisting through the winter.

DALIBARDA L.

D. repens L. FALSE VIOLET. Common in moist woods. Scarce, *Lucy*. Somewhat resembling a violet. Leaves roundish-cordate, crenate; flowers white; sepals spreading or reflexed in flower, erect in fruit. Spreads by creeping shoots. June.

GEUM L.

G. Canadense Jacq. WHITE AVENS. Common in moist woods and thickets. In foliage and flower much like a *Ranunculus*. June. (*G. album* Gmelin.)

G. Virginianum L. WHITE MARSH AVENS. Less common than the preceding, often growing with it, but usually in marshy ground. Petals yellowish white, shorter than the calyx. June.

G. strictum Ait. YELLOW AVENS. FIELD AVENS. BLACK-BUR. Common in fields, thickets and the borders of woods. Flowers bright yellow, resembling buttercups. June.

G. rivale L. PURPLE AVENS. WATER AVENS. Common in swamps. Flower nodding; sepals brownish-red; petals yellow, purplish outside; fruit in a globular head. An interesting plant. May. June.

WALDENSTEINIA WILLD.

W. fragarioides (Michx.) Tratt. BARREN STRAWBERRY. Common in open woods and thickets in dry or moist soil. In aspect somewhat like the strawberry. Flowers, several on a scape, bright yellow. May.

FRAGARIA L.

F. Virginiana Duchesne. WILD STRAWBERRY. Very common in fields and meadows. Achenes embedded in the receptacles. May.

F. Americana (Porter). Britton. ALPINE STRAWBERRY. Common, *Graves*. Not infrequent, *Clute*. Found in rocky upland woods. A native plant usually confused with the introduced *F. vesca* of Europe. Berries light red, pointed; achenes superficial on the receptacle. The variety *alba*, the "Indian strawberry," is occasionally found in rocky soil. Fruit white, leaves thicker and shining, *Graves*.

F. vesca L. GARDEN STRAWBERRY. Frequently escapes from cultivation. Distinguished from the preceding by its thicker, broader and more hairy leaflets; its rounder, blunter, deeper red berries, and its more robust habit. (*F. vesca* L.)

POTENTILLA L.

P. arguta Pursh. CINQUEFOIL. Frequent, *Graves*. Not common, *Clute*. Rare, *Lucy*. Elsewhere not noted. Found in fields. Leaves pinnate; leaflets five or more; flowers in cymes, with roundish, light-yellow petals. Apparently is increasing in numbers. June.

P. Monspeliensis L. NORWAY CINQUEFOIL. THREE-LEAVED CINQUEFOIL. Common in fields and waste places. An erect species with small yellow flowers. (*P. Norvegica* L.)

P. argentea L. SILVERY CINQUEFOIL. Common, except in Susquehanna county, where it is apparently missing. Found in fields, along roadsides and in other dry grounds. A pretty little plant, with trailing stems and small yellow flowers. Leaves bright green above, covered beneath with silvery white down. June–August.

P. Anserina L. SILVERWEED. GOOSE GRASS. In the "Cayuga Flora" this is recorded from Cayuta Lake. The plant is not uncommon on sandy shores throughout the central part of New York State, and should be searched for in other places within our limits.

P. Canadensis L. COMMON CINQUEFOIL. FIVE-FINGER. Very common in dry fields; spreading by long runners. Flowers bright yellow, on peduncles from the axils of the leaves. The petals in all our species of Cinquefoil are rather fugacious.

P. recta L. HIGH CINQUEFOIL. An escape from cultivation which has become abundantly naturalized and is fast spreading in parts of Broome county. Plant erect, a foot or more high; flowers in terminal cymes, light yellow. In meadows and along roadsides, *Clute*.

COMARUM L.

C. palustre L. MARSH CINQUEFOIL. Swamp at Elmira; scarce, *Lucy*. Mutton Hill pond; common, *Fenno*. Pond Brook and bog near Jarvis street, Binghamton; not common, *Clute*. Beebe's swamp and Cranberry marsh; common, *Graves*. Common about ponds at Preston, *Coville*. Cayuta Lake, *Hoy*. Not reported from Delaware county. Found only in boggy places.

A curious plant, in aspect like a Cinquefoil, with petals much smaller than the sepals. Stamens, styles, petals and upper surface of the calyx dark red. June. (*Potentilla palustris* Scop.)

AGRIMONIA L.

A. striata Michx. AGRIMONY. Common in thickets.

SANGUISORBA L.

S. Canadensis L. CANADIAN BURNET. In wet places. Willow Point, *Millspaugh.* Near Newtown creek at Horseheads; very scarce, *Lucy.* Rare, *Brown.* (*Poterium Canadense* Benth. & Hook.)

ROSA L.

R. blanda Ait. EARLY WILD ROSE. Not uncommon on rocky shores. Stems nearly unarmed. Petals red. June.

R. Carolina L. SWAMP ROSE. CAROLINA ROSE. DOG ROSE. Common in swamps where it often forms dense thickets, six feet or more in height. Branches scarcely armed. The red fruit is very noticeable after the leaves have fallen. June.

R. lucida Ehrh. SHINING WILD ROSE. Occasional in moist soil, *Graves.* Not common, *Clute.*

R. humilis Marsh. DWARF ROSE. Common on dry or rocky slopes, forming low thickets. Flowers pink. Our most plentiful species. Often flowers in Autumn. This species so nearly resembles *lucida* and *blanda* that they are easily confused by the novice. June.

R. rubiginosa L. EGLANTINE. SWEET BRIAR. Naturalized along roadsides. Well-known from its aromatic foliage.

PYRUS L.

P. coronaria L. WILD CRAB-APPLE. Generally distributed, but not very plentiful, in thickets, fence-rows and along streams. A small tree, with thorny branches. Flowers large, pink, very fragrant; fruit about one inch in diameter, yellowish-green, sour, greasy to the touch, fragrant. Our most fragrant wild flower. May.

P. malus L. APPLE. In woodlands, thickets and fence rows. Well naturalized. Fruit of inferior size, hard and gnarly, but not unpleasant to the taste when mellowed.

ARONIA Pers.

A. nigra (Willd.) Britton. CHOKEBERRY. CHOKE HUCKLEBERRY. Very common in swamps and bogs, often forming thickets of considerable extent. A shrub, six to eight feet or more high. Flowers small, numerous, in cymes, white or pinkish; fruit black. (*Pyrus arbutifolia*, var. *melanocarpa* Hook.)

SORBUS L.

S. Americana Marsh. AMERICAN MOUNTAIN ASH. Found wild in Susquehanna county; occasionally cultivated elsewhere. Distinguished from the Mountain Ash of Europe, which is also cultivated, by its smaller habit, greener leaflets, less downy leaves and buds, and smaller berries.

CRATÆGUS L.

C. coccinea L. SCARLET-FRUITED THORN. WHITE THORN. CRIMSON HAW. Common in woods, thickets and pastures. Fruit bright red. The variety *macracantha* is reported from the Chemung valley by Dr. Lucy.

C. tomentosa L. BLACK THORN. Rare. Near Elmira and Corning, *Lucy*. Ross Park, *Millspaugh*.

C. punctata Jacq. DOTTED-FRUITED THORN. Not uncommon in fields and thickets. Fruit rather large, red or yellow, occasionally dotted with white.

C. Crus-galli L. COCK-SPUR THORN. Rather common in woods and thickets. Not reported west of Tioga county. Thorns four inches or more in length. April. May.

AMELANCHIER Medic.

A. Canadensis (L.) Medic. SHAD-BUSH. JUNEBERRY. SERVICEBERRY. Common in woods. A tree, sometimes of considerable size. Flowers white, in long racemes, appearing with or before the leaves; fruit red or purplish, sweet, edible. The blooming of the tree is popularly supposed to herald the approach of shad in the rivers. Apr. May.

A. Botryapium (L. f.) DC. Frequent, *Lucy*. Rare, *Clute*. Occasional, *Graves*.

SAXIFRAGACEÆ.

SAXIFRAGA L.

S. Virginiensis Michx. EARLY SAXIFRAGE. Common on moist, rocky banks. Rare in the northern part of Chenango county, *Coville*. Leaves radical, in a rosette; scape much branched at top; flowers white, numerous. One of our earliest flowers. April-June.

S. Pennsylvanica L. SWAMP SAXIFRAGE. Common in swamps and wet woodlands. Plant coarse and stout, much larger than the foregoing, which in general aspect it resembles. Flowers greenish. May.

TIARELLA L.

T. cordifolia L. FALSE MITRE-WORT. BISHOP'S-CAP. FEVER-WORT. Very common in damp, shaded woods and ravines. Spreads by leafy runners. Flowers, several, white, on a scape. The leaves when bruised, emit a sweetish odor. May.

MITELLA L.

M. diphylla L. MITRE-WORT. Common. Found with the preceding which it greatly resembles, but distinguished from it by its finely divided, white petals. Flowering stems bearing a pair of opposite leaves near the middle. In summer runners are produced on which the leaves are alternate. May.

M. nuda L. NAKED MITRE-WORT. Very rare, *Brown*. A search in our deep woods and cold ravines may reveal other stations for this species.

CHRYSOPLENIUM L.

C. Americanum Schwein.; Hook. GOLDEN SAXIFRAGE. WATER CARPET. Common. Found in wet woods and ditches, usually growing in water. Leaves small, roundish; flowers greenish, conspicuous only by reason of their orange-colored anthers. May.

PARNASSIA L.

P. Caroliniana Michx. GRASS OF PARNASSUS. Rare. Valley of the Chenango near Port Crane; base of Spanish hill, *Millspaugh*. Leaves smooth, thick, ovate; flowers on short scapes, white, veined with greenish. In localities just beyond our range this plant is very common on springy banks. It is probable that further search will prove it to be more plentiful with us. August.

HYDRANGEA L.

H. arborescens L. WILD HYDRANGEA. Frequent in the Chemung valley. Near Waverly; rare, *Graves, Millspaugh*. This plant is also quite common in Bradford county, Pa., always in situations with a northern exposure, in rocky soil near water, *Lucy*. Not reported from other parts of our range.

RIBES L.

R. Cynosbati L. WILD GOOSEBERRY. Common in thickets, fence rows and neglected fields. Stems armed with sharp prickles; fruit prickly, brownish-red when ripe, sweet, edible.

R. rotundifolium Michx. ROUND-LEAVED GOOSEBERRY. MOUNTAIN CURRANT. Infrequent, *Lucy*. Occasional, *Graves*. Common, *Hoy*. Found in rocky soil.

R. prostratum L'Her. FETID CURRANT. Not rare, *Graves*. Occasional, *Coville*. Very scarce, *Lucy*. Not found elsewhere. Prefers wet, rocky soil.

R. floridum L'Her. WILD BLACK CURRANT. Not uncommon in thickets, wet or dry. Flowers greenish-yellow.

R. aureum Pursh. MISSOURI OR BUFFALO CURRANT. GOLDEN CURRANT. An occasional escape, with bright yellow, fragrant flowers, and black or yellow fruit.

———— The European Currant (*R. rubrum*) belongs here. It may escape, but seldom persists long.

CRASSULACEÆ.

PENTHORUM L.

P. sedoides L. DITCH STONE-CROP. VIRGINIA STONE-CROP. Common in wet open places. Flowers in one-sided branches, usually without petals, their parts in fives, sixes or sevens. July.

SEDUM L.

S. acre L. MOSSY STONE-CROP. Found in several places, in rocky soil along roadsides, *Graves*. Rare, *Clute*. Scarce, *Fenno*.

S. ternatum Michx. THREE-LEAVED STONE-CROP. Infrequent, Franklin, *Hoy*. Occasional, *Brown*. Rare; probably an escape from cultivation, *Clute*.

S. Telephium L. LIVE-FOR-EVER. GARDEN ORPINE. Common in dry fields, along roadsides, etc., also in the vicinity of streams. Leaves fleshy; flowers in a compound cyme, white or purple, usually the latter. A well-known plant, abundantly naturalized. Does not flower freely within our limits, being spread chiefly by its tuberous roots. July.

DROSERACEÆ.

DROSERA L.

D. rotundifolia L. ROUND-LEAVED SUNDEW. Not common. Mutton-Hill Pond, *Fenno*. Pond Brook, *Clute*. Beebe's swamp; Butler's Lake, *Graves*. "The Vlai." near Oneonta, *Hoy*. Not reported from the Chemung valley. Found only in peaty bogs, but usually plentiful in such places. Leaves, roundish, on long petioles, in a rosette on the ground; flowers, white, in a coiled raceme which unrolls as the flowers open. The upper surface of the leaf is covered with red, glandular bristles, giving a reddish hue to the plant. July.

D. intermedia Hayne. NARROW-LEAVED SUNDEW. Rare. Round Pond, McDonough; Plymouth Pond, Brisbin Pond, *Coville*. Butler's Lake, Susquehanna county. Found by Messrs. Hoy and Graves in July, 1894. Has much the appearance of the preceding, except that the leaves are spatulate-oblong. (*D. intermedia* Hayne, var. *Americana* DC.)

HAMAMELIDACEÆ.

HAMAMELIS L.

H. Virginiana L. WITCH HAZEL. Common and well-known; in woods, thickets and fence-rows in any kind of soil. Blooms in late Autumn as its leaves are falling, and ripens its seeds during the succeeding twelve months. Flower-buds formed in July, opening in October. Flowers yellow, with a peculiar heavy odor, in clusters along the branches. Seeds, when ripe, expelled from the capsules with considerable force. A very conspicuous shrub in the leafless Autumn woods.

HALORRHAGIDACEÆ.

MYRIOPHYLLUM L.

M. spicatum L. Common in the Chenango river at Oxford, *Coville*.

CALLITRICHACEÆ.

CALLITRICHE L.

C. palustris L. WATER STAR-WORT. Not rare, *Graves*. Infrequent, *Clute*. Somewhat common, *Coville*. Occasional, *Fenno*. Not reported elsewhere. Found in slow streams, where it forms mats of green on the surface. Submersed leaves narrow; floating leaves spatulate, in rosettes. July. (*C. verna* L.)

LYTHRACEÆ.

ROTALA L.

R. ramosior (L.) Koehne. Not rare. Near the Susquehanna river in stagnant water, *Graves*.

DECODON J. F. Gmel.

D. verticillatus (L.) Ell.   Swamp Loose-strife.   Water Willow.   Not common.   Lakes six miles south of Apalachin. *Fenno.*   Pond Brook; Cutler's Pond, *Clute.*   Beebe's swamp; Butler Lake, *Graves.*   Northern end of Warn's Pond, *Coville.* Not noted elsewhere within our limits.   Grows in shallow water on the borders of lakes and ponds, and resembles a low, half-shrubby willow.   Flowers in clusters from the axils of the upper leaves, rose-purple.   August.

ONAGRACEÆ.

LUDWIGIA L.

L. palustris (L.) Ell.   Water Purslane.   Swamp near Oakland; not rare, *Graves.*   Along Baldwin creek and the Chemung river; frequent, *Lucy.*   Near Oxford, *Coville.*   Not reported from other places within our limits.

CHAMÆNERION Adans.

C. angustifolium (L.) Scop.   Great Willow-herb.   Fire-weed. Common along roadsides, river banks, and especially in newly cleared land.   Plant willow-like in appearance.   Flowers large, numerous, rose-purple, in a terminal raceme.   June-August. (*Epilobium angustifolium* L.)

EPILOBIUM L.

E. coloratum Muhl.; Willd.   Small Willow-herb.   Purple-veined Willow-herb.   Very common in wet, open places.   A weedy-looking plant, with inconspicuous flowers.

E. strictum Muhl.   Near Warn's Pond, Oxford, *Coville.*

E. palustre L.   Chenango valley near Binghamton; Mutton-Hill Pond, *Millspaugh.*

ONAGRA Adans.

O. biennis (L.) Scop.   Common Evening Primrose.   Common along roadsides and in other waste grounds.   Plant coarse and stout.   Flowers, large, yellow, in a terminal leafy spike, opening toward evening, occasionally fragrant.   (*Œnothera biennis* L.)

KNEIFFIA Spach.

K. pumila (L.) Spach. SMALL EVENING PRIMROSE. Common in fields in wet or dry soil. In aspect this plant is a good miniature of the preceding. June. July. (*Œnothera pumila* L.)

K. fruticosa (L) Raimann. SUNDROPS. Rather common, usually in dry soil. The plant is not shrubby, as its name would imply. (*Œnothera fructicosa* L.)

GAURA L.

G. biennis L. Common from Broome county west along river banks. Apparently absent north and east. A coarse, stout plant with wand-like spikes of rose-purple flowers that soon fall. July.

CIRCÆA L.

C. Lutetiana L. ENCHANTER'S NIGHTSHADE. Common in rich, moist woods. Flowers small, white, with their parts in twos, borne in a terminal raceme; fruit bur-like, covered with hooked bristles. Plant a foot or more in height.

C. alpina L. LOW ENCHANTER'S NIGHTSHADE. Common. Found in ravines and other cool, moist spots. Plant much resembling the preceding.

CUCURBITACEÆ.

SICYOS L.

S. angulatus L. ONE-SEEDED BUR-CUCUMBER. Common along streams, except in the northeastern part of our range. Stem branching, often thirty feet in length, climbing over trees and bushes. Staminate flowers greenish-white, in racemes; pistillate clustered in the axils of the leaves; fruit small, covered with sharp, bristly prickles. Often cultivated for shade. July, Sept

MICRAMPELIS Raf.

M. lobata (Michx.) Greene. CLIMBING CUCUMBER. BALSAM APPLE. FOUR-SEEDED CUCUMBER. More common than the preceding, for which it is easily mistaken when not in flower or fruit; grows in the same places. Stem seldom branching. Staminate flowers white, in long racemes; pistillate inconspicuous, succeeded by spongy, cucumber-like fruit, containing four seeds. Occasionally cultivated. July. August. (*Echinocystis lobata* T. & G.)

AIZOACEÆ.

MOLLUGO L.

M. verticillata L. CARPET-WEED. INDIAN CHICKWEED. An abundant weed in cultivated fields, along railroads and on sandy river shores. Not reported northeast of Broome county. Plant much branched, flat on the ground, forming circular patches. Leaves spatulate in whorls at the joints; flowers inconspicuous.

UMBELLIFERÆ.

DAUCUS L.

D. Carota L. WILD CARROT. BIRD'S-NEST. QUEEN ANNE'S LACE. An abundant and well-known weed in fields. Flowers white, or occasionally pink, in flat-topped, lace-like umbels.

ANGELICA L.

A. villosa (Walt.) B. S. P. WOOD ANGELICA. HAIRY ANGELICA. Common in woods. (*A. hirsuta* Muhl.)

A. atropurpurea L. PURPLE-STALKED ANGELICA. Abundant along streams. Often eight feet high, with a purplish, hollow stem, ample leaves and large, globular umbels. Our largest Umbelwort.

HERACLEUM L.

H. lanatum Michx. COW PARSNIP. Common in low grounds. Flowers white, in large umbels, very conspicuous when in bloom. June.

PASTINACA L.

P. sativa L. WILD PARSNIP. A very common and troublesome weed along roadsides, river banks and all waste places. Flowers yellow. Our plant is supposed to be the wild form of the cultivated parsnip, and to be poisonous. It is doubtful if this latter opinion is correct.

THASPIUM Nutt.

T. trifoliatum aureum (Nutt.) Britton. MEADOW PARSNIP. GOLDEN ALEXANDERS. Very common in meadows, thickets and open woodlands; the radical leaves resembling those of the marsh marigold. (*T. aureum* Nutt.)

T. barbinode (Michx.) Nutt. GOLDEN ALEXANDERS. Frequent in meadows, *Graves*. In alluvial soil, *Lucy*. Near Waverly, *Millspaugh*. Elsewhere not reported, although it doubtless occurs.

PIMPINELLA L.

P. integerrima (L.) A. Gray. GOLDEN ALEXANDERS. Common from Susquehanna county west. Not reported northward. Found in rocky ground. Leaflets entire.

DERINGA Adans.

D. Canadensis (L.) Kuntze. HONEWORT. Frequent in moist, rich soil, especially in woodlands. (*Cryptotænia Canadensis* DC.)

SIUM L.

S. cicutæfolium J. F. Gmel. WATER PARSNIP. Not uncommon throughout our range in marshy places.

ZIZIA Koch.

Z. aurea (L.) Koch. GOLDEN ALEXANDERS. Tolerably common from Susquehanna county west. Not noted northward. Earlier flowering than our species of *Thaspium*. Rays of the umbel longer and more numerous, *Graves*.

Z. cordata (Walt.) DC. HEART-LEAVED ZIZIA. Common. Found in the same places as the preceding and has the same range. The radical leaves are heart-shaped or nearly round, long petioled and blunt-toothed, rarely lobed, *Graves*.

CARUM L.

C. Carui L. CARAWAY. Becoming common along roadsides, about dooryards, etc. Easily mistaken for the wild carrot.

CICUTA L.

C. maculata L. WATER HEMLOCK. MUSQUASH ROOT. SPOTTED COWBANE. BEAVER POISON. Common in damp situations. A coarse plant, with purple-streaked stems. Root deadly poison.

C. bulbifera L. BULB-BEARING WATER HEMLOCK. Common along streams, ponds and swamps. Rare, *Lucy*. Slender; divisions of the leaflets long and narrow. In the axils of the leaves are borne numerous bulbs, which drop from the plant and are carried to other places by the water. In spring the bulbs are often found floating on the borders of ponds, with leaves and roots well developed.

CONIUM L.

C. maculata L. POISON HEMLOCK. Common in meadows and waste grounds. A tall, coarse plant, with purple stem, whose foliage exhales a disagreeable odor when bruised. Root deadly poison.

CHÆROPHYLLUM L.

C. procumbens (L.) Crantz. WILD CHERVIL. Rare; in rich, moist soil. River bank, town of Ashland, *Lucy*.

OSMORRHIZA RAF.

O. Claytoni (Michx.) B. S. P. HAIRY SWEET CICELY. Common in rich, moist woods. Leaves hairy, style very short, conical; root bitter. (*O. brevistylis* DC.)

O. longistylis (Torr.) DC. SMOOTH SWEET CICELY. Found in the same situations as the preceding, but not so common. Not reported north of Broome county. Easily confounded with the other species. Distinguished by its nearly smooth, often purplish stem, its less divided leaves and longer styles, the latter one-twelfth of an inch long. Root branching; with a pleasant, anise-like flavor.

ERIGENIA NUTT.

E. bulbosa (Michx.) Nutt. HARBINGER-OF-SPRING. PEPPER-AND-SALT. Plentiful on first island north of Noyes' island—the only station, *Millspaugh*.

HYDROCOTYLE L.

H. Americana L. WATER PENNY-WORT. Very common in wet places, generally in shade. An interesting little plant, with roundish, crenately-lobed leaves whose veins are more prominent on the upper than on the under surface. Umbels axiliary and almost sessile.

SANICULA L.

S. Marylandica L. SANICLE. BLACK SNAKEROOT. Common in rich, moist woodlands.

ARALIACEÆ.

ARALIA L.

A. spinosa L. HERCULES CLUB. ANGELICA TREE. Rare. Roadside at Oakland, *Graves*. Near Churchill's Lake, *Clute*. Apalachin, in cultivation, *Fenno*. Mouth of Cayuta creek, *Millspaugh*. Not found elsewhere. A singular looking shrub, eight to ten feet high, seldom branching; leaves in a tuft at the top, very large, decompound; trunk and stems prickly.

A. racemosa L. SPIGNET. SPIKENARD. PETTYMORREL. WHITEROOT. Common in moist, shady places, especially in ravines. Roots large, spicy-aromatic, edible.

A. hispida Vent. BRISTLY SARSAPARILLA. WILD ELDER. Apparrently restricted to the more elevated parts of our range. Rare in Chemung county, more common in Steuben county, *Lucy*. Thompson's marsh, abundant, *Clute*. Glenwood, *Millspaugh*. Frequent, *Graves*. Franklin, *Hoy*. Oxford, *Coville*. Elsewhere not reported. In clearings, old fields, etc , especially about old stumps and stone-heaps. Stem at base, shrubby, beset with sharp prickles. Umbels many, globous.

A. nudicaulis L. WILD SARSAPARILLA. Common throughout in woodlands. Rootstock long, aromatic, edible; leaf solitary, decompound; scape bearing three or more globular umbels. May. June.

PANAX L.

P. quinquefolium L. GINSENG. Rare, except in the eastern part of our range where it is yet comparatively common. Found in rich woods. In aspect like the preceding species, with the peduncle bearing the single umbel rising from the midst of the leaves. Root large, spindle-shaped, edible. This plant is gradually disappearing, owing to the demand for its roots which are still held in some estimation as a drug. Easily cultivated. (*Aralia quinquefolia* Decsne & Planch.)

P. trifolium L. GROUND-NUT. DWARF GINSENG. Common in rich woods and thickets. Stem from a globular, edible tuber, rather deep in the ground. At the top of the stem is a whorl of three compound leaves, from the center of which rises the single umbel of pure-white flowers. Barren and fertile flowers on different plants. Occasionally fragrant. May. (*Aralia trifolia* Decsne & Planch.)

CORNACEÆ.

CORNUS L.

C. Canadensis L. LOW CORNEL. BUNCH-BERRY. Common in damp, shady places. Stem a few inches high, bearing a whorl of about six leaves and a single head of small flowers, which with their broad, white, four-leaved involucre are easily mistaken for a single blossom. Berries red. May.

C. florida L. FLOWERING DOGWOOD. Common, especially in upland woods. A well known tree with flowers like the preceding. Involucres much larger, pure white or pinkish, obcordate, very noticeable when expanded in the vernal woods.

C. circinata L'Her. GREEN OSIER. ROUND-LEAVED CORNEL or DOGWOOD. Not uncommon in woods and thickets. Bark green with warty spots. June.

C. Amonum Mill. KINNIKINNIK. SILKY CORNEL. Common in low grounds, borders of swamps, etc. Stems purplish; branchlets and under surface of the leaves downy; fruit pale blue. (*C. sericea* L.)

C. stolonifera Michx. RED OSIER or DOGWOOD. Abundant along streams, borders of swamps, etc., forming tangled thickets. Stems red, very noticeable in winter. Fruit white. June.

C. candidissima Marsh. PANICLED CORNEL. Frequent in thickets in wet or dry soil. Much branched; bark gray, branchlets chestnut-colored; flowers many, white, in conical cymes; fruit white. June. (*C. paniculata* L'Her.)

C. alternifolia L. f. ALTERNATE-LEAVED CORNEL. Common in open woods and copses. Leaves alternate, clustered at the ends of the spreading, warty, greenish branches. Flowers numerous, pale buff color; fruit deep blue, on reddish stalks. June.

——The fruit of the Cornels, especially that of *florida*, *Amomum* and *stolonifera*, is much liked by thrushes, sparrows, woodpeckers and other birds.

NYSSA L.

N. aquatica L. PEPPERIDGE. SOUR GUM. BLACK GUM. TUPELO. Common in rich, moist soil. Scarce, *Lucy*. Not reported north of Broome and Susquehanna counties. A tree often of considerable size. Leaves oblong, firm, shining, turning bright crimson in autumn; drupes bluish black; branches usually gnarled and twisted; wood soft, but very hard to split because of the interlacing fibres. (*N. sylvatica* Marsh.)

CAPRIFOLIACEÆ.

SAMBUCUS L.

S. Canadensis L. COMMON ELDER. Abundant in fence rows, along roadsides, and especially in the vicinity of water. Flowers numerous, white, in flat cymes; berries black, edible. Well known.

S. pubens Michx. RED-BERRIED ELDER. Common, especially in moist, rocky woods. Blooms much earlier than the preceding; berries smaller, scarlet, ripe in June. Reputed to be poisonous. The fruit is greedily eaten by many of our smaller birds. Very noticeable when in fruit. (*S. racemosa* L.)

VIBURNUM L.

V. alnifolium Marsh. HOBBLE BUSH. WITCH HOPPLE. AMERICAN WAYFARING TREE. WILD HYDRANGEA. Common in cool, moist woods. Leaves large, roundish-ovate; flowers small, white, in a flat cyme, the outer ones sterile, with much larger corollas; drupes dark red, edible. The long, slender branches often root at their extremities. May. (*V. lantanoides* Michx.)

V. Opulus L.   High Cranberry. Cranberry Tree Found sparingly throughout our range except in the valley of the Chemung river. Usually occurs in thickets, especially in damp places. Leaves three-lobed; flowers as in the preceding, fruit red, acid, sometimes used as a substitute for the cranberry. The Guelder Rose of the gardens is a variety of this.

V. acerifolium L. Dockmackie. Maple-leaved Viburnum. Arrow-wood. Common in rocky woods. A well marked species. Flowers small, greenish-white; fruit purplish. June.

V. pubescens (Ait.) Pursh. Downy Arrow-wood. Not common. Reported from Susquehanna county west. Frequent, *Lucy*. Much resembles *V. dentatum*, but is distinguished from it by its shorter petioles and the presence of awn-like stipules. June.

V. dentatum L. Arrow-wood. More common than the preceding. Found on the borders of swamps and in other low grounds. Young branches long, straight and slender; fruit, purple.

V. Cassinoides L. Withe-rod. Tolerably common from Susquehanna and Broome counties west. Elsewhere not reported. Found in low grounds.

V. Lentago L. Nanny-berry. Sweet Viburnum. Sheep-berry. Common throughout in low grounds. A shrub or small tree, well known for its dark-blue, edible fruit, ripe in September.

V. prunifolium L. Black Haw. Sloe. Frequent in fence-rows and along roadsides, *Graves*. Often confounded with the preceding, from which it is distinguished by the rather obtuse and finally serrate leaves, and slightly margined petioles. It should occur in other parts of our range.

TRIOSTEUM L.

T. perfoliatum L. Fever-wort. Horse-gentian. Tinker's-weed. Wild Coffee. Not very common. Usually found on the borders of woods. Flowers in the axils of the leaves, brownish red, succeeded by orange-colored, fleshy fruits. May.

LINNÆA L.

L. borealis L. Twin flower. Rare. In woods, wet or dry. Hill south of Susquehanna, *Graves*. Killawog, *Clute*. Oxford, *Coville*. Tioga Centre, *Fenno*. Corning; also Bradford county, Pa., *Lucy.* Sexsmith lake, *Hoy.* Stems extensively creeping and sending up at intervals short branches; leaves roundish-ovate, crenate; flowers pink, in pairs on a common peduncle, nodding, fragrant. In general appearance this plant is much like *Mitchella.* Easily cultivated. May.

SYMPHORICARPOS Juss.

S. racemosus Michx. SNOW BERRY. Occasionally found throughout our range, along roadsides, etc. Probably an escape. The pure white berries make the plant noticeable in autumn.

LONICERA L.

L. ciliata Muhl. FLY HONEYSUCKLE. VERNAL HONEYSUCKLE. Common in rocky woods and ravines. A pretty little shrub, bearing numerous straw-colored flowers in pairs in early spring; fruit, bright red. April. May.

L. dioica L. SMALL or GLAUCOUS HONEYSUCKLE. Not uncommon in open woods and thickets. Upper leaves connate, glaucous, especially underneath; corollas dull red, hairy within. June. (*L. glauca* Hill.)

DIERVILLA Mœnch.

D. Diervilla (L.) MacM. BUSH HONEYSUCKLE. Abundant along fence-rows and roadsides, in thickets and dry woods. A low shrub, with oblong-ovate leaves and yellowish corollas, the styles and stamens much exserted. June. (*D. trifida* Mœnch.)

RUBIACEÆ.

HOUSTONIA L.

H. cærulea L. BLUETS. INNOCENTS. FORGET-ME-NOTS. DWARF PINKS. Abundant. Found in woods, fields, and especially in meadows, where they often form a dense carpet over large areas. Well known. Within our limits this plant has been found blooming in every month of the year except January and February. May.

CEPHALANTHUS L.

C. occidentalis L. BUTTON BUSH. Common on the borders of swamps, forming low thickets. Flowers, tubular, with exserted styles, assembled into close globular heads. July. Aug.

MITCHELLA L.

M. repens L. PARTRIDGE-BERRY. SQUAW-BERRY. TEA-BERRY. EYE-BERRY. Abundant in damp woodlands. Stems trailing; leaves roundish, entire; flowers in pairs, funnel-shaped, white or pink, hairy within, fragrant; berry bright red, composed of the united ovaries and marked with the lobes of the two calyxes. Fruit well flavored but scarcely edible on account of the hard, bony seeds. June.

GALIUM L.

G. Aparine L. GOOSE-GRASS. CLEAVERS. Very common in damp, shaded places. Leaves about eight in a whorl; stem covered with hooked prickles pointing downward; flowers very small, white. June.

G. pilosum Ait. HAIRY CLEAVERS. Rare, *Graves*. Town of Ashland, *Lucy*.

G. circæzans Michx. WILD LIQUORICE. Tolerably common in woods. Not reported from the Chenango valley. The leaves are said to have a sweetish taste like liquorice.

G. lanceolatum Torr. WILD LIQUORICE. LANCE-LEAVED BED-STRAW. More common than the preceding, which it much resembles. Found in woods throughout our range.

G. boreale L. NORTHERN BEDSTRAW. Very common on banks, wet or dry, in sun or shade. Stems erect, a foot or more high; flowers white, minute, numerous, in a thryse-like panicle at the top of the stem. A pretty and decorative plant, very conspicuous when in bloom. June.

G. trifidum L. SMALL BEDSTRAW. DYER'S CLEAVERS. Not uncommon in swamps and bogs throughout.

G. asprellum Michx. ROUGH BEDSTRAW. Common in swamps and low grounds.

G. triflorum Michx. SWEET-SCENTED BEDSTRAW. Common in damp woodlands. Foliage sweet-scented in drying.

VALERIANACEÆ.

VALERIANELLA Poll.

V. chenopodiifolia (Pursh) DC. LAMB'S LETTUCE. Common in the river meadows of Broome county, *Clute; Millspaugh*.

V. radiata (L.) Dufr. LAMB'S LETTUCE. Common in a meadow north of Oxford, *Coville*. Barton; occasional, *Fenno*.

V. Woodsiana patellaria (Sulliv.) A. Gray. Common. Found with the preceding, *Coville*.

DIPSACEÆ.

DIPSACUS L.

D. sylvestris Huds. TEASEL. Common along roadsides and in other waste grounds. A curious, thistle-like plant, with long, close-set, prickly heads of purplish flowers. Apparently spreading. July.

SCABIOSA L.

S. australis Wulf. The only known station for this plant within our limits is in the vicinity of Lisle, where it was first noticed growing along the railroad. It has since spread six miles south and is increasing its territory. Leaves much as in *Dipsacus*, but not prickly; flowers, in roundish heads, blue or purplish, *Clute*.

COMPOSITÆ.

EUPATORIUM L.

E. purpureum L. TRUMPET-WEED. JOE-PYE WEED. PURPLE BONESET. Common in low grounds. A tall, coarse weed, with purplish flowers.

E. perfoliatum L. BONESET. THOROUGHWORT. Abundant in meadows and low grounds. Well known and valued for its medicinal qualities. Aug.

E. ageratoides L. f. WHITE SNAKEROOT. Common on shady banks and the borders of woods and thickets. Aug.

SOLIDAGO L.

S. squarrosa Muhl. BIG GOLDENROD. Common on dry hill-tops and along upland roadsides. Not reported north and east of Broome county. Easily distinguished from all others by its large flowers and squarrose scales. Aug. Sept.

S. cæsia L. SLENDER GOLDENROD. BLUE-STEMMED GOLDENROD. Common on dry banks and along bushy roadsides. Flowers clustered in the axils of the leaves. An elegant species. Aug. Sept.

S. flexicaulis L. BROAD-LEAVED GOLDENROD. Not uncommon in moist, shady places. Flowers clustered in the axils of the leaves. Remarkable for its broad, cauline leaves. Aug. (*S. latifolia* L.)

S. bicolor L. Mant. WHITE GOLDENROD. SILVERROD. Common in upland woods and thickets and along roadsides. Flowers white or nearly so. Our only white goldenrod. The so-called variety, *concolor*, with yellow rays, is occasionally found. Aug.

S. patula Muhl.; Willd. Along Baldwin creek, Chemung county; swamp near Cohocton, Steuben county; infrequent, *Lucy*. Barton, *Fenno*. Swamps along the Unadilla; not common, *Brown*.

S. rugosa Mill. ROUGH GOLDENROD. Abundant in fence-rows, thickets and the borders of woods. Stem mostly branching at top, the branches terminated by one-sided panicles of flowers. Leaves very rough, those of the branches much smaller than the rest. A homely but noticeable species. Aug.

S. ulmifolia Muhl.; Willd. ELM-LEAVED GOLDENROD. Found on dry, warm slopes. Mountain House narrows, town of Big Flats; banks east of Elmira, infrequent. *Lucy*.

S. arguta Ait. GOLDENROD. Not uncommon in wet or dry soil. In swampy places the leaves are often rough above.

S. juncea Ait. EARLY GOLDENROD. Very common in fields and pastures. Leaves rather thick. Our earliest species, often blooming before the hay is cut. July.

S. serotina Ait. SMOOTH GOLDENROD. Common along fences, roadsides, and in fields. Stem, smooth, purplish; leaves, large; inflorescence, rather small. A handsome plant. Aug.

S. serotina gigantea (Ait.) A. Gray. Not uncommon. Found usually in wet ground. Distinguished from the species in being taller, with green stems, thinner and lighter colored leaves, and larger inflorescence. Aug.

S. Canadensis L. CANADIAN GOLDENROD. Common along fences, in thickets, etc. Stem, tall hairy; leaves pubescent below, rough above; inflorescence rather large. Aug. Sept.

S. nemoralis Ait. GRAY GOLDENROD. LOW GOLDENROD. Abundant in dry, sterile soil. Stem and leaves grayish; inflorescence large for the plant, and quite conspicuous. Aug.

EUTHAMIA NUTT.

E graminifolia (L.) Nutt. NARROW-LEAVED GOLDENROD. Very common in moist ground, but found in dry soil as well. Leaves linear; flowers, rather small, in corymbs. Aug. (*Solidago lanceolata* L.)

SERICOCARPUS NEES.

S. asteroides (L.) B. S. P. WHITE-TOPPED ASTER. Dry hill, Lanesboro; rather scarce, *Graves*. Sullivan hill, and elsewhere; frequent, *Lucy*. (*S. conyzoides* Nees.)

ASTER L.

A. divaricatus L. CORYMBED ASTER. EARLY ASTER. Common in woods and thickets. (*A. corymbosus* Ait.)

A. macrophyllus L. LARGE-LEAVED ASTER. Common. Frequently found with the preceding.

A. Novæ-Angliæ L. NEW ENGLAND ASTER. Common in damp meadows, fields, and along fence-rows. Rare, *Graves*. Stem stout, hairy; leaves numerous, lanceolate, pubescent; heads, large, with numerous violet-purple rays. Our most handsome species. Grows well in cultivation. The so-called variety with pink rays occurs frequently. Aug. Sept.

A. patens Ait. SPREADING ASTER. Somewhat rare in dry, rich woods. Not reported from the valleys of the Chemung and Chenango.

A. undulatus L. WAVY-LEAVED ASTER. Frequent in hilly woods. Heads middle-size; rays blue. Remarkable for the variability of its leaves. Sept.

A. cordifolius L. HEART-LEAVED ASTER. Common in open woods, along fence-rows, roadsides, and in thickets. Aug. Sept.

A. cordifolius polycephalus Porter. MANY-HEADED ASTER. Occasional; usually found with the type, *Graves; Lucy*. More robust and branched, leaves smaller, panicle ample, heads usually smaller and very numerous.

A. sagittæfolius Wedem.; Willd. ARROW-LEAVED ASTER. Not frequent. East Hill road near Elmira, *Lucy*.

A. Lowrieanus Porter. LOWRIE'S ASTER. Not uncommon. Until recently this species has been confused with *A. cordifolius*, but is easily distinguished by its broadly margined petioles, and thickish, leathery, perfectly smooth leaves. In a fresh state the leaves seem greasy to the touch. It will probably be found throughout our range.

A. Lowrieanus lanceolatus Porter. Occasional in Susquehanna and Chemung counties, *Graves; Lucy*. This variety doubtless has the same range as the type, and probably occurs elsewhere within our range.

A. lævis L. SMOOTH ASTER. Common in the borders of woodlands and in thickets. Not reported from the Chenango valley. Leaves very smooth, ovate-lanceolate, the upper clasping by a heart-shaped base; heads numerous, rays rather long, blue. A very showy species. Sept.

A. ericoides L. HEATH-LIKE ASTER. Chemung narrows; Mountain House narrows—the only stations; plentiful, *Lucy*.

A. vimineus Lam. SLENDER-STEMMED ASTER. Somewhat rare. Not reported from the Chenango valley.

A. lateriflorus (L.) Britton. STARVED ASTER. Very common in dry soil; fields, roadsides and thickets. Heads, numerous, very small, borne on one-sided branches; rays white. Sept. (*A. diffusus* Ait.)

A. paniculatus Lam. PANICLED ASTER. Common, especially in low grounds. Sept.

A. prenanthoides Muhl.; Willd. SWAMP ASTER. Rather common in moist, open woodlands, and along the borders of streams.

A. prenanthoides porrectifolius Porter. Rare. In moist places along wood-roads, *Graves*. Specimens in Lafayette College Herbarium.

A. puniceus L. Rough stemmed Aster. Swamp Aster. Common in wet grounds. Stem tall, rough; heads rather large. Sometimes mistaken for *A. Novæ-Angliæ*, but the rays of this are much lighter in color. Sept.

A. umbellatus Mill. Umbelled Aster. Not plentiful. River banks and swamps, *Graves*.

A. infirmus Michx. Double-bristled Aster. Sullivan Hill; frequent, *Lucy*.

A. acuminatus Michx. Pointed-leaved Aster. Plentiful in rich woods. Not reported from the Chemung valley.

ERIGERON L.

E. Canadensis L. Fleabane. Mare's-tail. Horse-weed. Butter-weed. Very common in waste places. Stem tall, much branched at top, with numerous small heads of flowers. Aug.

E. annuus (L.) Pers. Tall Daisy. White-weed. Sweet Scabious. Common in fields, meadows and along roadsides.

E. ramosus (Walt.) B. S. P. Daisy Fleabane. Rough-stemmed Fleabane. Common. Found in the same localities as the preceding. (*E. strigosus* Muhl.)

E. pulchellus Michx. Robin's Plantain. Common in grassy fields. Heads rather large, the rays numerous, pink-purple. May. (*E. bellidifolius* Muhl.)

E. Philadelphicus L. Common Fleabane. Pink Fleabane. Plentiful in moist, grassy places. Heads medium sized, with very many pink rays scarcely wider than threads. May. June.

ANTENNARIA Gærtn.

A. plantaginifolia (L.) Richards. Mouse-ear Plantain. Plantain-leaved Everlasting. Very common in woodlands, thickets and old fields, always growing in patches. Spreads by runners and offsets. Well known. April.

A. margaritacea (L.) Hook. Pearly Everlasting. Indian Tobacco. Common in old fields and occasionally in thickets. Leaves linear-lanceolate, grayish; involucral scales many, pearly-white. Aug. (*Anaphalis margaritacea* Benth. & Hook.)

GNAPHALIUM L.

G. obtusifolium L. CUD-WEED. COMMON EVERLASTING. Plentiful on dry hills and in old fields. Heads conical, with yellowish flowers. The leaves are fragrant. Aug. (*G. polycephalum* Michx.)

G. decurrens Ives. EVERLASTING. Found with the preceding. Not reported from the Chemung valley. May be distinguished from the others by its decurrent leaves. Aug.

G. uliginosum L. LOW CUDWEED. Common. Usually found along roadsides where water has stood; also in other wet places. Heads very small; leaves fragrant. Aug.

INULA L.

I. Helenium L. ELECAMPANE. Common, especially in low grounds and along roadsides. A well-known plant. The root has the smell of camphor. Aug.

POLYMNIA L.

P. Canadensis L. LEAF-CUP. Rare. South Mountain, *Millspaugh*. Apalachin, *Fenno*. Chemung Narrows, *Lucy*. Sheshequin, Pa., *Graves*.

AMBROSIA L.

A. trifida L. RIVER RAG-WEED. GREAT RAG-WEED. THREE-LOBED RAG-WEED. Abundant along streams, where its tall stems form dense brakes. The so-called variety, *integrifolia*, which is lower, with usually entire leaves, occurs occasionally with the type.

A. artemisiæfolia L. RAG-WEED. ROMAN WORMWOOD. HOG-WEED. BITTER-WEED. Abundant in all waste places and many cultivated grounds. Plants that are entirely pistillate are sometimes found. This plant has the reputation of being the cause of hay-fever.

XANTHIUM L.

X. strumarium L. COCKLE-BUR. Rare. Found about barnyards and pastures, *Graves; Coville; Barbour*.

X. Canadense Mill. CLOT-BUR. COCKLE-BUR. Common along roadsides, railways, river banks and other waste places. A coarse weed that appears to be increasing in numbers.

HELIOPSIS Pers.

H. helianthoides (L.) B. S. P. OX-EYE. Common on the banks of rivers and streams. A tall, coarse plant with the aspect of a wild sunflower. Aug. (*H. lævis* Pers.)

H. scabra Dunal. ROUGH OX-EYE. Rare. Island in Chemung river, *Lucy*.

RUDBECKIA L.

R. laciniata L. TALL CONE-FLOWER. Common in wet, usually shady grounds. Stem tall; heads large; rays about eight, yellow, spreading or drooping. Aug.

R. hirta L. CONE-FLOWER. YELLOW DAISY. BROWN-EYED SUSAN. BULL'S EYE DAISY. Very common in fields, meadows and pastures. Heads large; rays bright yellow; disk-flowers brownish purple. The rays occasionally have brownish markings at the base. Monstrous forms are not uncommon in which from two to five heads are united. July. Aug.

HELIANTHUS L.

H. annuus L. COMMON SUNFLOWER. The sunflower of the gardens is commonly found as an escape about rubbish heaps, but seldom attains the size of the cultivated plant and does not increase in numbers.

H. divaricatus L. WOOD SUNFLOWER. Abundant in thickets, along roadsides and the banks of streams. Leaves ovate-lanceolate, sessile, opposite, thick, rough; heads large, though small for the genus; rays bright yellow. No scrubby hillside is too dry for this plant. Our earliest species to bloom. July. Aug.

H. strumosus L. SUNFLOWER. Frequent along the Chemung river and tributary streams, *Lucy*. Susquehanna county; not common, *Graves*. Binghamton, *Clute*. Apalachin, *Fenno*.

H. decapetalus L. TEN-RAYED SUNFLOWER. RIVER SUNFLOWER. Common along streams, but often found in dry ground. Stem tall; leaves rather large, ovate, contracted into margined petioles. Rays about ten.

H. tuberosus L. JERUSALEM ARTICHOKE. Not uncommon in waste places, dry or wet.

BIDENS L.

B. frondosa L. SPANISH NEEDLES. BEGGAR-LICE. STICK-TIGHTS. PITCHFORKS. DEVIL'S NEEDLES. Abundant, usually in damp ground. Well known. Aug.

B. connata Muhl.; Willd. SWAMP BEGGAR-TICKS. BUR MARIGOLD. Common along ditches and in swamps. Aug.

B. cernua L. SMALLER BUR MARIGOLD. Common in low, moist places.

B. lævis (L.) B. S. P. LARGE BUR MARIGOLD. Common in swamps and along streams and ditches. Aug. Sept. (*B. chrysanthemoides* Michx.)

B. bipinnata L. SPANISH NEEDLES. In moist soil; not common; Oakland, *Graves*.

GALINSOGA R. & P.

G. parviflora Cav. A rather common weed in fields and gardens about Binghamton, *Clute*. Not reported elsewhere. Leaves ovate, acute; heads small and inconspicuous; disk-flowers yellow; rays white, usually three-lobed. Aug.–Oct.

HELENIUM L.

H. autumnale L. SNEEZE-WEED. Common on the banks of rivers and streams, and in swamps. Plant tall; leaves lanceolate; disk globular; rays numerous, bright yellow. A showy plant that thrives well in cultivation. Sept. Oct.

ANTHEMIS L.

A. Cotula L. MAY-WEED. DOG FENNEL. STINKING CHAMOMILE. Common in damp places along roadsides, and about dooryards. Well known.

ACHILLEA L.

A. Millefolium L. YARROW. MILFOIL. Common and well known. In fields and pastures. The form with rose colored rays is frequently found.

CHRYSANTHEMUM L.

C. Leucanthemum L. WHITE DAISY. OX-EYE. WHITE-WEED. Very abundant, especially in fields and meadows, where it often forms the principal vegetation. Disliked by the farmer.

———*C. Parthenium* and *C. Balsamita* are occasionally found as escapes, but do not persist.

TANACETUM L.

T. vulgare L. TANSY. Common along roadsides, along streams and about old buildings. Well known. The fresh leaves are considered a sure preventive against the depredations of the carpet beetle.

———The Wormwood, Mugwort and Southernwood (*Artemisia Absinthium*, *A. vulgaris* and *A. Abrotanum*), belong here. All are occasionally found as escapes, but seldom exist long in the wild state.

TUSSILAGO L.

T. Farfara L. COLTSFOOT. Common, except in the Chemung valley where it is rare. Found along roadsides, and especially on the banks of streams. Flowers before the leaves in early spring, like dandelions, on scaly scapes; leaves, large, cordate, angular. This plant is a lover of clay soils, and its presence is usually an indication of such. Mar. April.

SENECIO L.

S. aureus L. GOLDEN RAGWORT. GROUNDSEL. SQUAW-WEED. PIUNKUM. Common on the borders of swamps and along streams. Stems simple or branched above; heads numerous; rays golden yellow. May.

CACALIA L.

C. suaveolens L. INDIAN PLANTAIN. River banks near Oxford, *Coville*. Apalachin; rare, *Fenno*.

C. atriplicifolia L. PALE INDIAN PLANTAIN. Very rare. A single plant collected at Wellsburg in August, 1874, *Lucy*.

ERECHTITES RAF.

E. hieracifolia (L.) Raf.; DC. FIREWEED. BUTTER-WEED. Common in waste places, especially in grounds that have been recently burned over. A coarse, ill-smelling weed, with cylindric involucres and rayless flowers. Aug.

ARCTIUM L.

A. Lappa L. BURDOCK. A common and well-known plant about dwellings, in fields, etc. The varieties *minus* and *majus* occasionally occur.

CARDUUS L.

C. lanceolatus L. COMMON THISTLE. Plentiful in waste places, usually in good soil. (*Cnicus lanceolatus* Hoffm.)

C. discolor (Muhl.) Nutt. TALL THISTLE. Not rare. Found in low grounds. This species is easily confused with *C. muticus*, but may be distinguished by its leaves which are white woolly beneath. (*Cnicus altissimus* Willd., var. *discolor* A. Gray.)

C. muticus (Michx.) Pers. SWAMP THISTLE. Abundant along streams and in swamps. Stem very tall, scarcely prickly; heads medium sized, purple. This species when in bloom attracts numerous ruby-throated humming-birds. Aug. (*Cnicus muticus* Pursh.)

C. odoratus (Muhl.) Porter. PASTURE THISTLE. Not uncommon in fields and pastures. Stem low, heads few, large, fragrant. An ornamental species. (*Cnicus pumilus* Torr.)

C. arvensis (L.) Rob. CANADA THISTLE. CURSED THISTLE. Abundant in fields, meadows and waste grounds. Heads small, purple. A troublesome pest, hard to eradicate. A white-flowered form is occasionally found. (*Cnicus arvensis* Hoffm).

CICHORIACEÆ.

CICHORIUM L.

C. Intybus L. CHICKORY. SUCCORY. Becoming common along roadsides and in other waste grounds. Heads large, bright blue. The root is used in adulterating coffee.

TRAGOPOGON L.

T. porrifolius L. SALSIFY. VEGETABLE OYSTER. An occasional escape from cultivation found along roadsides and about dwellings. Flowers purplish.

T. pratensis L. GOAT'S BEARD. GO-TO-BED-AT-NOON. Becoming common in fields, along roadsides and in waste places. Flowers bright yellow, closing by mid-day. Often mistaken for the preceding species.

HIERACIUM L.

H. pilosella L. MOUSE-EAR HAWKWEED. Montrose, 1878, *C. H. Peck*. Vicinity of New Milford, *Graves; Clute*. Leaves clustered at the base of the stem, hairy; heads several, large, bright yellow. Spreads by runners. A species recently introduced from Europe. May. June.

H. aurantiacum L. ORANGE HAWKWEED. DEVIL'S PAINT-BRUSH. Common in fields. Leaves nearly all radical, densely hairy; heads clustered at the top of the stem, crimson, with orange centers. This plant was first known from our region about ten years ago, and is becoming plentiful in many places.

H. Canadense Michx. CANADA HAWKWEED. Frequent in dry woods. Not reported from the Chenango valley. Stem simple, leafy; heads corymbed, rather large. Aug.

H. paniculatum L. PANICLED HAWKWEED. Not uncommon in dry, open woods

H. venosum L. RATTLESNAKE-WEED. Common in dry woods and thickets. Leaves all radical, veined with purple; stem, very slender; heads small, yellow. June.

H. scabrum Michx. ROUGH HAWKWEED. Common in woods. Stem rather stout, leafy; heads large.

PRENANTHES L.

P. alba L. RATTLESNAKE-ROOT. WHITE LETTUCE. DROP FLOWER. LION'S FOOT. Common in thickets and the borders of woods. Stem, tall; heads pendulus, in corymbs, dingy white; pappus brown. Aug.

P. serpentaria Pursh. LION'S FOOT. GALL-OF-THE-EARTH. About as common as the preceding, which it much resembles. Not reported from the Chenango valley. Flowers white, yellowish or purplish; pappus straw-color. Aug.

P. altissimus L. SLENDER DROP-FLOWER. Not uncommon in rich, moist woods. Aug. Sept.

TARAXACUM HALL.

T. Taraxacum (L.) Karst. DANDELION. Common everywhere. One of the first flowers to bloom in spring and the last to close in autumn. Has been found blooming ten months in the year in our region. (*T. officinale* Weber.)

LACTUCA L.

L. Scariola L. PRICKLY LETTUCE. Not common. A plant of recent introduction that seems to be spreading. Reported only from Susquehanna, Broome and Chemung counties.

L. Canadensis L. WILD LETTUCE. TRUMPET MILKWEED. Com- in thickets and fence-rows. A tall, coarse plant, with many small heads of yellow flowers. July.

L. villosa Jacq. BLUE LETTUCE. Not uncommon in dry ground. Not reported from the Chenango valley. Flowers bluish. (*L. acuminata* Gray.)

L. spicata (Lam.) A. S. Hitchc. WHITE LETTUCE. Common in waste places. Flowers bluish-white. (*L. leucophæa* Gray.)

SONCHUS L.

S. oleraceus L. COMMON SOW THISTLE. Plentiful in waste places, especially along railways and roadsides.

S. asper (L.) All. PRICKLY SOW THISTLE. Common in the same situations as the preceding.

LOBELIACEÆ.

LOBELIA L.

L. cardinalis L. CARDINAL FLOWER. Common on islands and the shores of our larger streams. Very rarely found along the smaller streams. Flowers, large, in a terminal spike, brilliant red. The plant spreads by offsets in late summer. This species is very unequally distributed. It occurs in great quantities at some points, but generally the plants are found singly. July. Aug.

L. syphilitica L. GREAT BLUE LOBELIA. Common in moist places, especially the borders of swamps and wet meadows. Stem, rather tall; flowers, large, blue. Aug.

L. spicata Lam. SPIKED LOBELIA. Common in the Susquehanna valley, rare in the Chemung, and not reported from the Chenango. Found in fields and pastures, dry or wet. Spike, long and slender; flowers, small, blue. July.

L. Kalmii L.   KALM'S LOBELIA.   Rare.   Base of South Mountain, *Millspaugh*.   Pond Brook, *Clute*.

L. inflata L.   INDIAN TOBACCO.   Common in old fields and on hills throughout our range.   Flowers small, blue; seed-pods inflated.   Plant poisonous, but once in great repute as a medicine. July.

CAMPANULACEÆ.

LEGOUZIA DURAND.

L. perfoliata (L.) Britton.   VENUS' LOOKING-GLASS.   Common in dry fields, along roadsides and in cultivated grounds.   Stem simple, or sometimes branched; flowers rather large, deep bluish-purple, in the axils of the small ovate leaves.   June. (*Specularia perfoliata* A. DC.)

CAMPANULA L.

C. rapunculoides L.   BELLFLOWER.   Common along roadsides, and in the vicinity of old dwellings.   An escape from cultivation.

C. rotundifolia L.   HARE-BELL. BLUE-BELL.   Rather rare.   Found on ledges and rocky hillsides.   Hills north of Susquehanna, *Graves*.   Chenango Forks; Willow Point, *Millspaugh*.   Near Waverly, *Barbour; Millspaugh*.   Mt. Zoar Hill, *Lucy*.   Ledges along the Susquehanna from Towanda to Tunkhannock, *Clute*. Not reported from the upper Chenango valley.   Stem slender; leaves linear; flowers rather large, drooping, deep blue.   The early leaves are roundish.   July. Aug.

C. aparinoides Pursh.   MARSH BELLFLOWER.   Common in wet, grassy places.   Stem slender, rough; flowers nearly white, bell-shaped.   Plant with much the habit of a Galium.   July.

C. Americana L.   TALL BELLFLOWER.   Very rare.   Mountain House Narrows, west of Elmira, *Lucy*.   Waverly, *Graves*.

ERICACEÆ.

GAYLUSSACIA H. B. K.

G. resinosa (Ait.) T. & G.   BLACK HUCKLEBERRY.   Common in rocky woods and on the borders of swamps.   Fruit black, without bloom; seeds hard.

VACCINIUM L.

V. stamineum L. DEER-BERRY. SQUAW HUCKLEBERRY. Common on dry, bushy hillsides. A low shrub, with numerous white flowers, succeeded by large, green berries, bitter to the taste. When in bloom the plant is very noticeable. May.

V. Pennsylvanicum Lam. DWARF BLUEBERRY. EARLY BLUEBERRY. Common in dry, hilly woods and pastures, and along fence-rows. Flowers, reddish-white; berries blue, sweet. The earliest of our blueberries to ripen. May.

V. vacillans Kalm; Torr. LOW BLUEBERRY. Not uncommon. Found with the preceding species which it much resembles. May.

V. corymbosum L. HIGH HUCKLEBERRY. SWAMP HUCKLEBERRY. Common in swamps, but often found in dry soil. Rare in the Chemung valley, *Lucy*. A good-sized shrub, which in favorable situations forms dense thickets. In our territory this species produces the bulk of the huckleberries sent to market. May.

V. atrococcum (A. Gray.) Heller. BLACK HUCKLEBERRY. Dry ground beyond Bear Swamp, near Susquehanna, *Graves*. Fruit black without bloom. (*V. corymbosum*, var. *atrococcum* Gray.)

SCHOLLERA ROTH.

S. Oxycoccus (L.) Roth. SMALL CRANBERRY. Not very common. Found only in bogs, usually growing in sphagnum. Mutton-Hill pond; bogs north of Barton, *Fenno*. Bog near Jarvis street, Binghamton; Pond Brook, *Clute*. Cranberry marsh; Beebe's swamp, *Graves*. "The Vlai," near Oneonta, *Hoy*. Near Oxford, *Coville*. Not reported from the Chemung valley. Stems slender, creeping or trailing; leaves very small, evergreen; flowers rose-colored, nodding; berries red, often gathered for market. June. (*Vaccinium Oxycoccus* L.)

S. macrocarpa (Ait.) Britton. LARGE CRANBERRY. AMERICAN CRANBERRY. Less common than the preceding, and found in the same places. Resembles *S. Oxycoccus*, but is larger. This species is the one usually cultivated and yields the bulk of our cranberries. June. (*Vaccinium macrocarpon* Ait.)

CHIOGENES Salisb.

C. hispidula (L.) T. & G. Creeping Snowberry. Running Birch. Rare. Found in cold, wet woods and swamps. Barton, *Fenno*. Newark Valley, *Barbour*. Pond Brook, *Clute*. Swamp near New Milford, *Graves*. Near Oxford, *Coville*. Unadilla Forks, *Brown*. A slender, trailing vine, with tiny, evergreen leaves, white flowers and snow-white berries. The whole plant has a pleasant spicy flavor like wintergreen or birch. May. (*C. serpyllifolia* Salisb.)

EPIGÆA L.

E. repens L. Trailing Arbutus. Mayflower. Ground Laurel. Common in open deciduous woods, on bushy hillsides and on knolls in swamps. Among the first of our vernal flowers. The demand for the fragrant, rose-colored blossoms is causing this species to be slowly exterminated in the vicinity of our cities and towns. It is very difficult to make this plant grow in cultivation. Flower buds for the succeeding spring are formed in September, and the plant occasionally blooms in October. April, May.

GAULTHERIA L.

G. procumbens L. Wintergreen. Aromatic Wintergreen. Checkerberry. Teaberry. Common in open woods and thickets, in wet or dry soil. In many dry, deciduous woods the undergrowth consists almost entirely of this species. Berries red, edible, sometimes gathered for the markets. Leaves evergreen, with a pungent, spicy flavor, which renders them in great demand during May and June when the flavor is strongest.

ANDROMEDA L.

A. Polifolia L. Marsh Rosemary. Common in the eastern part of our range; not reported west of Broome county. Found in bogs and swamps. Leaves narrow, evergreen, white beneath, the margins revolute; flowers, in umbel-like clusters, pinkish-white, globose-cylindrical. A pretty little undershrub, forming dense patches in sphagnum bogs. May.

XOLISMA Raf.; Britton.

X. ligustrina (L.) Britton. Swamp Andromeda. Common, except in the western part of our range. Rare, *Lucy*. Found in swamps, also on dry hills near the site of exsiccated peat swamps, *Coville*. A low, straggling shrub, with white flowers in panicles, like diminutive blueberry blossoms. June. (*Andromeda ligustrina* Muhl.)

CHAMÆDAPHNE Moench.

C. calyculata (L.) Moench. LEATHER-LEAF. Common. Rare in the Chemung valley. Found in bogs and marshes, where it forms low thickets ofter acres in extent and grows so luxuriantly that other plants are excluded. Stems about two feet high; leaves oblong, nearly evergreen, in autumn turning russet and remaining on the plant until spring; flowers white, urn-shaped, in leafy racemes. Flower-buds of this and the preceding species formed in autumn and quite prominent in winter. May. (*Cassandra calyculata* Don.)

KALMIA L.

K. latifolia L. MOUNTAIN LAUREL. CALICO-BUSH. SPOONWOOD. Abundant on rocky hillsides and often on the borders of swamps. Rare in the Chenango valley, and not common in Broome county. Shrub sometimes ten feet high, but usually much lower; leaves thick, shining, evergreen; flowers large, saucer-shaped, angular, pink or white, in terminal corymbs. Reputed to be poisonous. One of our showiest wildflowers. June.

K. angustifolia L. NARROW-LEAVED LAUREL. SHEEP LAUREL. LAMB-KILL. Found only in the eastern part of our territory, but there abundant, growing in swamps and bogs. Not noted from Broome county west. A lower shrub than the preceding, with narrower leaves, and much smaller, deep pink flowers, in axillary fascicles, on the young shoots. June.

K. glauca Ait. GLAUCOUS LAUREL. PALE LAUREL. SMOOTH LAUREL. Rare. Found in swamps and on boggy shores. Butler and Goodrich Lakes, *Hoy*. Churchill Lake, *Graves*. Bog near Union, *Clute*. North of Barton, *Fenno*. The only stations. Stems low; branches two-edged; leaves narrow, white-glaucous beneath, the margins revolute; flowers middle-sized, lilac-purple. May, June.

AZALEA L.

A. nudiflora L. MAYFLOWER. PINKSTER. SWAMP PINK. HONEYSUCKLE. PINK AZALEA. Common in woods and thickets, wet or dry. Flowers large, in clusters, varying from deep pink to nearly white, appearing with or before the leaves; stamens much exserted. The blossoms are so fragrant that the whole wood is scented when they bloom. The flowering parts frequently become changed to greenish-white excrescences, covered with white bloom and containing an insiped juice relished by children. May, June. (*Rhododendron nudiflora* Torr.)

A. Canescens Michx. MOUNTAIN AZALIA. Rare. Pond Brook, *Clute*. The only station. This is much farther north than it usually occurs. Differs from the preceding in having wider and shorter soft-canescent leaves, glandular pedicels, etc. (*Rhododendron nudiflora* Torr.)

RHODORA L.

R. Canadensis L. RHODORA. Rare. Beebe's swamp; swamp in North Jackson, *Graves*. The only stations. A shrub much resembling the preceding, with blossoms of rose purple. May. (*Rhododendron Rhodora* Don.)

RHODODENDRON L.

R. maximum L. RHODODENDRON. DEER LAUREL. BIG LAUREL. Rare. Carmalt Lake, *Fenno*. Bear swamp, *Graves*. Swamp near Unadilla Forks, *Brown*. Swamp near Cincinnatus, *L. H. Dewey*. Not noted elsewhere. A tall shrub or small tree. Leaves often ten inches long and four inches wide, very thick, evergreen; flowers large, bell-shaped, spotted inside with green, yellow and purple, in many-flowered corymbs. Just south of our range this species is very abundant. July.

LEDUM L.

L. Grœnlandicum Œder. Common in peat bogs near Oxford, *Coville*. Not reported elsewhere. Our territory is nearly the southern limit of this species. A thorough search in all our cold bogs may reveal a few more stations for this plant (*L. latifolium* Ait.)

PYROLACEÆ.

CHIMAPHILA PURSH.

C. umbellata (L.) Nutt. PRINCE'S PINE. PIPSISSEWA. Common in rather dry, rich woods. Stems low; leaves mostly near the top, lanceolate, evergreen; flowers in terminal umbels, flesh-colored, the anthers violet. The whole plant has a bitter taste, and is valued in medicine. June.

C. maculata (L.) Pursh. SPOTTED WINTERGREEN. Rare. Midland woods, Oxford, *Coville*. Mt. Prospect, *Millspaugh*. East of Cohocton, on the edge of the swamp, *Lucy*. The only stations. Much resembles the preceding and grows in the same situations. Leaves, mottled with white.

MONESES SALISB.

M. uniflora (L.) A. Gray. ONE-FLOWERED PYROLA. Very rare. The Tower woods, Oxford, 1886, *Coville*. (*M. grandiflora* Salisb.)

PYROLA L.

P. secunda L. ONE-SIDED PYROLA. LOW SHINLEAF. Common in dryish woods, frequently in the shade of coniferous trees. Leaves small, ovate; flowers greenish-white, in one-sided racemes; styles straight. June. July.

P. chlorantha Sw. FALSE WINTERGREEN. Common in open, upland woods. Not reported from the Chenango valley. Flowers greenish-white, larger than in the preceding, nodding, the styles declined and curved. Our commonest Pyrola. June. July.

P. elliptica Nutt. SHINLEAF. Common in dry, rich, shaded soil. Resembles *P. chlorantha*, but with white, fragrant flowers, which open later. July.

P. rotundifolia L. ROUND-LEAVED WINTERGREEN. Common in rich woods. Flowers, nodding, white, fragrant, in a long raceme. July.

MONOTROPACEÆ.

PTEROSPORA NUTT.

P. Andromedea Nutt. PINE DROPS. Very rare. A single specimen found under a white pine in upland woods, southeast of Fitch's Bridge, Chemung county, July, 1892, *Lucy*.

MONOTROPA L.

M. uniflora L. INDIAN PIPE. CORPSE PLANT. PINE-SAP. Common in the eastern part of our range; infrequent or rare west of

Broome county. Found in wet or dry soil in shade. Flowers usually several together, white, nodding, on scaly stems from a ball of fibrous rootlets; capsules erect in fruit. The flowers are usually plentiful after the first hard rain in July.

HYPOPITYS ADANS.

H. Hypopitys (L.) Small. PINE-SAP. FALSE BEECHDROPS. BIRD'S-NEST. Rare. Found under pine trees. Unadilla Forks, *Brown*. Oxford, *Coville*. Franklin, *Hoy*. Brandt; Bear Swamp, *Graves*. Mt. Prospect, *Clute*. Ross Park, *Millspaugh*. Not reported west of Broome county. Stems scaly, tawny or reddish, bearing a cluster of several half-nodding, usually fragrant flowers. Petals usually red; essential organs yellow or orange. August. (*Monotropa Hypopitys* L.)

PRIMULACEÆ.

TRIENTALIS L.

T. Americana Pursh. CHICKWEED-WINTERGREEN. STAR-FLOWER. Common in low, rich woods. Stem low, with a whorl of lance-olate leaves at top; flowers two or three on slender pedicels, white. May.

STEIRONEMA RAF.

S. ciliatum (L.) Baudo. CILIATE LOOSESTRIFE. Common in thickets, and along streams. Leaves on long, fringed petioles. June. July.

LYSIMACHIA L.

L. quadrifolia L. FOUR-LEAVED LOOSESTRIFE. WHORLED LOOSE-STRIFE. Common in thickets and open woods. Leaves two to six, in whorls on the simple stem; flowers yellow, on slender peduncles from the axils of the leaves. June.

L. terrestris (L.) B. S. P. RACEMED LOOSESTRIFE. Very common on the borders of ponds, in swamps and on river shores. Flowers numerous, yellow, on slender pedicels, in a long terminal raceme. After blooming the plant produces an abundance of bulblets in the axils of the leaves. June-Aug. (*L. stricta* Ait.)

L. Nummularia L. MONEYWORT. Common and troublesome in lawns. Formerly cultivated for its bright yellow flowers, but now naturalized in many places. Stems creeping and rooting; leaves roundish; flowers axillary. July. Aug.

NAUMBURGIA MŒNCH.

N. thyrsiflora (L.) Duby. TUFTED LOOSESTRIFE. Rare. Found only in bogs. Pond Brook, *Clute*. Brisben Pond, *Coville*. Butler's Lake, *Graves*. Near Waverly, *Millspaugh*. Cinnamon Lake, *Lucy*. Near Barton, *Fenno*. The only stations. Leaves lanceolate, opposite; flowers small, numerous in capitate spikes, from the axils of a few of the upper leaves. May. June. (*Lysimachia thyrsiflora* L.)

ANAGALLIS L.

A. arvensis L. SCARLET PIMPERNEL. Very rare. Near Elmira. *Lucy*. Unadilla Forks, *Brown*. The only stations.

OLEACEÆ.

FRAXINUS L.

F. Americana L. WHITE ASH. Common in rich, moist woods throughout.

F. Pennsylvanica Marsh. RED ASH. PUBESCENT-STEMMED ASH. Less common than the preceding. Not reported from the Chenango valley. Found in low grounds. Known by its velvety-pubescent young shoots and leaf stalks. (*F. pubescens* Lam.)

F. nigra Marsh. BLACK ASH. WATER ASH. Common in moist woods and swamps. The tough wood splits readily and is extensively used in basket making. The bruised leaves have the odor of the elder.

LIGUSTRUM L.

L. vulgare L. PRIVET. PRIM. Often planted for hedges and occasionally escapes, *Graves; Fenno*.

APOCYNACEÆ.

VINCA L.

V. minor L. PERIWINKLE. RUNNING MYRTLE. A common and well-known trailing shrub, in cemeteries and about old dwellings; often found escaped along roadsides. Leaves evergreen; flowers purple.

APOCYNUM L.

A. androsæmifolium L. DOG-BANE. MILKWEED. Very common in thickets, fence-rows and open woods. Plant one to two feet high, spreading; flowers numerous, bell-shaped, white, striped with purple inside. June. July.

A. Cannabinum L. INDIAN HEMP. Common on gravelly or sandy shores. Restricted to the banks of the rivers and larger streams. Flowers small, greenish-white. July. Aug.

ASCLEPIADACEÆ.

ASCLEPIAS L.

A. tuberosa L. BUTTERFLY-WEED. PLEURISY-ROOT. ORANGE MILKWEED. Common in the western part of our range. Pond Brook; rare, *Clute*. Not reported north, east or south of Broome county. Found in dry fields. Stems leafy, about two feet high; umbels numerous, terminal, forming a large corymb of bright, orange-yellow flowers. A handsome species when in bloom, easily distinguished at a distance from the goldenrods, with which it grows. Aug.

A. incarnata L. SWAMP MILKWEED. Common in low grounds. Stems tall, bearing at top numerous erect umbels of small, rose-purple flowers. July.

A. Syriaca L. COMMON MILKWEED. SILK-WEED. Very common and well known in dry fields, and along roadsides. Stems tall, stout; leaves oblong; flowers in large umbels, greenish or purplish-white, on stout peduncles from the axils of a few of the upper leaves, juice abundant, milky. The young shoots are extensively used as a pot-herb. July. (*A. Cornuti* Decaisne.)

A. exaltata (L.) Muhl. POKE MILKWEED. Common in open woodlands and thickets, oftenest in moist soil. Much resembles the preceding. July. (*A. phytolaccoides* Pursh.)

A. quadrifolia Jacq. FOUR-LEAVED MILKWEED. Common in dry open woods. Stems one to two feet long; leaves few, mostly in whorls of four; umbels few-flowered; flowers tinged with pink, slightly fragrant. June.

GENTIANACEÆ.

GENTIANA L.

G. crinita Frœl. FRINGED GENTIAN. Hill near Elmira; very rare, *Lucy.* Vicinity of Susquehanna; not rare, *Graves.* Glenwood; Pond Brook; plentiful at the latter station, *Clute.* Sidney, *Hoy.* Not reported elsewhere. Found usually in low grounds. Flowers two inches long; petals four, deep blue, fringed. A beautiful species, blooming in late autumn.

G. quinquefolia L. FIVE-FLOWERED GENTIAN. GALL-OF-THE-EARTH. Not uncommon in open, rocky places, usually in damp soil. Sept. (*G. quinqueflora* Lam.)

G. Andrewsii Griseb. CLOSED GENTIAN. Common along streams in open ground and in thickets. Stems one to two feet high; leaves opposite; flowers clustered in the axils of the leaves, an inch or more long, dark blue. Corolla inflated, but opening only slightly at top. A form with pure white flowers is often found with the type. Aug., Sept.

MENYANTHES L.

M. trifoliata L. BUCK-BEAN. Not uncommon in suitable places. Found in deep cold bogs. Leaves trifoliate, alternate; flowers in racemes, flesh-colored or white, the upper surface of the petals bearded. The common name is doubtless a corruption of "bog-bean." May.

POLEMONIACEÆ.

PHLOX L.

P. divaricata L. COMMON PHLOX. BLUE PHLOX. Occasional. Found in damp, open places. Stem about a foot high; flowers in corymbs, lilac or blue. May.

P. subulata L. MOUNTAIN PINK. MOSS PINK. GROUND PINK. Tolerably common in the western part of our range. Not reported from the Chenango valley, Broome and Susquehanna counties. Franklin, Sidney; common, *Hoy.* Found on warm, dry, usually rocky, slopes. A low moss-like plant, occurring in dense masses; in spring covered with many small pink flowers.

———*Phlox paniculata*, the Sweet William of the gardens, is occasionally found as an escape. It does not persist.

POLEMONIUM L.

P. reptans L. JACOB'S LADDER. Found sparingly throughout our range. Frequent, *Lucy*. Found on river banks. A handsome plant frequently cultivated.

P. Van-Bruntiæ Britton. JACOB'S LADDER. GREEK VALERIAN. *Coville; Hoy.* "Swamps about the source of the Susquehanna, N. Y."—*Grays Manual*, 5th ed. (*P. cœruleum* L.)

HYDROPHYLLACEÆ.

HYDROPHYLLUM L.

H. Virginicum L. WATER-LEAF. BURR-FLOWER. Common in shades, along roadsides and river banks. Corollas varying from white to blue. June.

H. Canadense L. WATER-LEAF. Rather rare. Reported from their localities by *Coville*, *Fenno* and *Lucy*.

BORRAGINACEÆ.

CYNOGLOSSUM L.

C. officinale L. HOUND'S-TONGUE. Common along roadsides and in waste places. Corolla a peculiar dull red, occasionally white. A well-known weed. June.

C. Virginicum L. WILD COMFREY. Somewhat rare. Found in open woods. Flowers pale blue.

LAPPULA MŒNCH.

L. Virginiana (L.) Greene. STICK-SEED. BURR-SEED. BEGGAR'S LICE. Not uncommon in thickets and waste places. Flowers small, white; fruit covered with many barbed prickles. (*Echinospermum Virginicum* Lehm.)

MERTENSIA ROTH.

M. Virginica (L.) DC. BLUE BELLS. SMOOTH LUNGWORT. VIRGINIA COWSLIP. Common along the banks of rivers and streams. Plant a foot or more high; leaves large, obovate; flowers many, in a terminal raceme, trumpet-shaped; corolla pink in the bud, deep blue in flower, or occasionally white. A beautiful plant, easily cultivated. May.

MYOSOTIS L.

M. laxa Lehm. FORGET-ME-NOT. Common on river banks, along ditches, and in swamps. Flowers numerous, small, light blue. June–Aug.

LITHOSPERMUM L.

L. arvense L. WHEAT THIEF. CORN GROMWELL. Somewhat rare. Reported by *Coville, Clute, Lucy* and *Fenno.*

ONOSMODIUM MICHX.

O. Carolinianum (Lam.) A. DC. FALSE GROMWELL. Rare. Near Harrington's ford; the only station, *Lucy.*

SYMPHYTUM L.

S. officinale L. COMFREY. Found occasionally throughout our range. An escape from gardens, but apparently well naturalized.

LYCOPSIS L.

L. arvensis L. SMALL BUGLOSS. Rare. Eldridge Park, Elmira, *Lucy.* Along the railroad, Susquehanna, *Graves.* Tioga Center, *Fenno.*

ECHIUM L.

E. vulgare L. VIPER'S BUGLOSS. BLUE-WEED. Otsego county, becoming common in fields, *Brown.* Broome county, not infrequent along roadsides and waste grounds, *Clute.* Elmira, along the railway, *Lucy.* Elsewhere not reported. Stem about two feet high, bristly; flowers numerous in crowded, axillary, recurved spikes; corollas rather large, blue. A showy plant when in bloom.

CONVOLVULACEÆ.

CONVOLVULUS L.

C. spithamæus L. LOW BIND-WEED. Common in dry soil; along upland roadsides and in thickets. Stem seldom more than a foot high; flowers white, funnel-shaped, about two inches long. June.

C. Sepium L.   HEDGE BIND-WEED. RUTLAND BEAUTY. WILD MORNING-GLORY. VIRGINIA CREEPER. Common in damp soil, along streams and the borders of fields. Occasionally found away from the river bottoms in dry soil. Stems several feet long, twining about bushes, etc. Flowers numerous, like the preceding, white, tinged with pink. Often cultivated.

C. repens L. LOW HEDGE BIND-WEED. Not common. Reported by *Graves*, *Clute* and *Lucy*. Differs from the preceding, which it closely resembles, by its prostrate, usually pubescent, stems, narrowly hastate or cordate leaves and obtuse bracts. (*C. sepium*, var *repens* Gray.)

C. arvensis L. FIELD BIND-WEED. Rare. Apalachin, D., L. & W. R. R. tracks, east of station, *Fenno*. Along Riverside Drive, Binghamton, *Clute*. The only stations. Leaves ovate-triangular, basal lobes acute; flowers three-quarters of an inch long, bell-shaped, white. July.

——*Ipomœa purpurea*, the common morning-glory, belongs here. It is often found growing about rubbish-heaps, but does not persist.

CUSCUTACEÆ.

CUSCUTA L.

C. Epithymum Murr. CLOVER DODDER. Very rare. Borders of a clover-field, *Graves*.

C. Gronovii Willd, R. & S. COMMON DODDER. LOVE VINE. Very common in damp, shady grounds. Stems bright orange, leafless, twining and parasitic on coarse herbs and shrubs. The golden-rod is most frequently the host plant, but others, as the blackberry and red maple, are occasionally attacked. Flowers small, white.

SOLANACEÆ.

SOLANUM L.

S. Dulcamara L. BITTERSWEET. WOODY NIGHTSHADE. Common, especially in damp soils. Flowers purplish-blue, anthers yellow; berries bright red. Reputed to be poisonous.

S. nigrum L. COMMON NIGHTSHADE. BLACK NIGHTSHADE. Rare. Big Island, Chemung river; waste ground, Elmira, *Lucy*. Binghamton, *Millspaugh*. Barton, *Fenno*. Oxford, *Coville*.

S. rostratum Dunal. BEAKED SOLANUM. Very rare. Cultivated ground, Susquehanna; apparently increasing, *Graves*. Exceedingly prickly, leaves pinnatifid. The original food plant of the potato bug.

PHYSALIS L.

P. pubescens L. GROUND CHERRY. STRAWBERRY TOMATO. Rare, *Lucy*. Infrequent, *Fenno*. Found in cultivated fields and waste places, not common; *Clute*. Elsewhere not reported. Corolla yellowish with brown-purple centre.

P. Virginiana Mill. GROUND CHERRY. Waverly, *Graves*. Oxford, *Coville*. Fitch's Bridge and along Baldwin Creek, *Lucy*. This plant is properly *P. heterophylla* of Nees.

———The apple of Peru (*Physalodes physalodes*), which belongs in this order, is sometimes found as an escape. It does not persist.

LYCIUM L.

L. vulgare (Ait. f.) Dunal. MATRIMONY VINE. In cultivation. Not uncommon about old dwellings, fences, etc. Occasionally escapes and shows a tendency to become naturalized.

DATURA L.

D. Stramonium L. JIMSON-WEED. THORN-APPLE. Rare. Found in waste grounds. Not reported from Delaware county. A rank, ill-scented weed, with showy, funnel-shaped, white blossoms, and globular, prickly, seed-capsules.

———The potato, egg-plant, tomato and tobacco belong to this family.

SCROPHULARIACEÆ.

VERBASCUM L.

V. Thapsus L. COMMON MULLEIN. Common in dry soil, along roadsides, etc. Well known. Its presence is usually an indication of sterile soil.

V. Blattaria L. Moth Mullein. Common, *Graves*. Not frequent, *Clute;* *Fenno*. Rare, *Lucy*. Owens Mills, Chemung county, scarce, *Barbour*. Not common, *Coville*. Not reported from Delaware county. Found in pastures, old fields and waysides. Stem tall, comparatively slender, branching above; flowers about an inch across, on slender pedicels, cream color or white, occasionally purplish; anthers unequal, purple. July.

LINARIA Juss.

L. Linaria (L.) Karst. Toad Flax. Butter-and-eggs. Jacob's Ladder. Ramstead. Abundant in dry fields and along roadsides. Flowers bright yellow in a dense raceme. Well known, and considered a difficult weed to kill. (*L. vulgaris* Mill.)

SCROPHULARIA L.

S. Marylandica L. Fig-wort. Common in fields, thickets and roadsides. (*S. nodosa* var. *Marilandica* Gray.)

CHELONE L.

C. glabra L. Turtle-head. Snake-head. Rheumatism-root. Common in open swamps and along rivers and streams throughout our range. Well known. Aug. Sept.

PENTSTEMON Soland.

P. hirsutus (L.) Willd. Beard-tongue. Abundant on dry banks and along upland roadsides. Stems low; flowers in open panicles, violet-purple, the sterile filament densely bearded. June. (*P. pubescens* Soland.)

P. Digitalis (Sweet) Nutt. Foxglove Beard-tongue. Tolerably common in rich soil, *Fenno*. Rare, *Clute*. Stem tall; flowers an inch long, abruptly dilated, white inclined to purplish. A beautiful species. June. (*P. lævigatus* var. *Digitalis* Gray.)

MIMULUS L.

M. ringens L. Monkey-flower. Common and well known. Found in swamps and on river shores. Flowers violet-purple, rarely lavender or white. July–Sept.

GRATIOLA L.

G. Virginiana L. HEDGE HYSSOP. Common throughout in muddy places. A low plant with an inconspicuous whitish or yellowish corolla.

ILYSANTHES RAF.

I. gratioloides (L.) Benth. FALSE PIMPERNEL. North side of Harrington's Island; rare, *Lucy*. River bank above Lanesboro; scarce, *Graves*. Mouth of Cayuta creek, *Millspaugh*. Near Oxford, *Coville*. Not reported from Broome and Delaware counties. Found on wet shores. (*I. riparia* Raf.)

LEPTANDRA NUTT.

L. Virginica (L.) Nutt. CULVER'S PHYSIC. CULVER'S-ROOT. Common in Broome and Tioga counties, infrequent in the Chemung valley. Not reported from the Chenango valley, nor in the Susquehanna valley east of Binghamton. Found on river flats and banks of the larger streams. Stem rather tall; leaves in whorls; flowers white, in spikes. At a little distance the plant is easily mistaken for *Cimicifuga* when in bloom. June, July. (*Veronica Virginica* L.)

VERONICA L.

V. Americana Schwein.; Benth. AMERICAN BROOKLIME. Frequent in ditches and slow rills. Stems spreading; flowers blue.

V. scutellata L. MARSH SPEEDWELL. SKULL-CAP. Tolerably common in wet places.

V. officinalis L. COMMON SPEEDWELL. Common on rather dry banks, roadsides and open woods. Stem prostrate, rooting at the lower joints.

V. serpyllifolia L. THYME-LEAVED SPEEDWELL. Common in lawns, fields, pastures and roadsides. May–July.

V. peregrina L. NECK-WEED. PURSLANE SPEEDWELL. Tolerably common except in the eastern part of our range. Not reported from Broome, Susquehanna and Delaware counties. Found in cultivated fields.

V. arvensis L. CORN SPEEDWELL. Common in the eastern, rare in the western part of our range. Found in cultivated grounds.

Stem erect, diffusely branched; flowers said to be whitish, but with us usually deep blue.

V. Byzantina (Sibth. & Smith.) B. S. P.   Rare.   In lawns in the village of Oxford, *Coville*.   (*V. Buxbaumii* Tenore.)

DASYSTOMA Raf.

D. Pedicularia (L.) Benth.   LOUSEWORT FOXGLOVE.   YELLOW FOXGLOVE.   Tolerably common in dry, open woods and thickets. Not reported north and east of Broome county.   Leaves resembling those of the common lousewort.   Flowers bell-shaped, bright yellow.   July.   (*Gerardia pedicularia* L.)

D. flava (L.) Wood.   DOWNY FALSE FOXGLOVE.   YELLOW FOXGLOVE.   Common in dry, open, upland woods.   Not reported north and east of Broome county.   Plant covered with a close down.   Flowers large, yellow.   July, Aug.   (*Gerardia flava* L.)

D. Virginica (L.) Britton.   SMOOTH FOXGLOVE.   OAK-LEAVED FOXGLOVE.   Not common, *Graves*.   Abundant, *Clute*.   Infrequent, *Fenno*.   Not reported from the Chemung and Chenango valleys, nor from Delaware county.   Found with the preceding. Lower leaves usually twice pinnatifid.   July, Aug.   (*Gerardia quercifolia* Pursh.)

GERARDIA L.

G. tenuifolia Vahl.   SLENDER GERARDIA.   PURPLE FOXGLOVE.   Tolerably common in Delaware, Susquehanna, Broome and Tioga counties on dry hillsides, in open copses, and along upland roadsides.   Rare in the Chemung, and not reported from the Chenango valley.   Stems spreading; leaves narrowly linear; corolla half an inch long, purple.   Aug., Sept.

CASTILLEJA Mutis; L, f.

C. coccinea (L.) Spreng.   SCARLET PAINTED-CUP.   Rare.   "Painted Post,"—*Cayuga Flora*.   Near Sayre, 1862, *Graves*.   The only stations.   Reported from Painted Post with yellow flowers.

PEDICULARIS L.

P. Canadensis L.   LOUSEWORT.   WOOD BETONY.   Common throughout in rather dry, open thickets.   Leaves finely pinnatifid.   A well known plant.

MELAMPYRUM L.

M. lineare Lam. COW-WHEAT. Common in dry, open woods and thickets. (*M. Americanum* Michx.)

OROBANCHACEÆ.

EPIPHEGUS Nutt.

E. Virginiana (L.) Bart. BEECH-DROPS. CANCER-ROOT. Not common. Found throughout our range under beech trees and parasitic on their roots. Stem branching, yellowish or purplish; leaves reduced to scales; flowers small, white marked with purple. Aug., Sept.

CONOPHOLIS Wallr.

C. Americana (L. f.) Wallr. SQUAW-ROOT. CANCER-ROOT. Very rare. Franklin, near road to Oneonta, *Hoy.* Oxford, *Coville.* Mt. Prospect, *Millspaugh.* The only stations.

THALESIA Raf.

T. uniflora (L.) Britton. ONE-FLOWERED CANCER-ROOT. Not common. Frequent, *Hoy.* Not reported from Broome county. Found in woods in rather dryish soil. Stem nearly subterranean; flowers on slender scapes. May. (*Aphyllon uniflorum* Gray.)

LENTIBULARIACEÆ.

UTRICULARIA L.

U. vulgaris L. COMMON BLADDERWORT. Reported from the eastern part of our range only. Common from Susquehanna and Broome counties north. Found floating in lakes, river-coves, sluggish streams and pools. Leaves with many capillary divisions, some of which bear bladders; flowers six or more, personate, yellow, borne above the water on a naked scape. July.

U. intermedia Hayne. Rare. Muddy borders of ponds near Oxford, *Coville.* Summit marsh, "*Cayuga Flora.*" The only stations.

———The butterwort (*Pinguicula vulgaris* L.) belongs to this family. It has been reported from Sidney, but specimens have not been seen, *Hoy.*

BIGNONIACEÆ.

CATALPA Scop.

C. Catalpa (L.) Karst. CATALPA. INDIAN BEAN. Not uncommon in cultivation and occasionally found as an escape. It is not quite hardy and many of the smaller branches are annually killed by the cold. Leaves cordate; flowers large, bell-shaped, white spotted with yellow and violet; pods often a foot long. (*C. Bignonioides* Walt.)

ACANTHACEÆ.

DIANTHERA L.

D. Americana L. WATER WILLOW. Found only in the Susquehanna river from the town of Tioga to our southern limits, *Fenno; Clute.* Occasionally grows on the river banks, but usually found in shallow water where it grows luxuriantly. July. Aug.

VERBENACEÆ.

VERBENA L.

V. urticæfolia L. WHITE VERVAIN. A common weed along roadsides and in waste places.

V. hastata L. BLUE VERVAIN. Common on the borders of swamps, in meadows and along roadsides. Usually in somewhat moist soil. Well known.

PHRYMA L.

P. Leptostachya L. LOP-SEED. Frequent throughout in woods and thickets. Flowers small, in a long terminal spike; fruit reflexed.

LABIATÆ.

TRICHOSTEMA L.

T. dichotomum L. BASTARD PENNYROYAL. Plentiful on the river flats opposite Apalachin, *Fenno.* Elsewhere not reported.

TEUCRIUM L.

T. Canadense L. WOOD SAGE. GERMANDER. Common in thickets on river banks and in low grounds. Not reported north of Broome county.

COLLINSONIA L.

C. Canadensis L. HORSE-BALM. RICH-WEED. STONE-ROOT. Common in moist woods and thickets. Flowers and foliage lemon-scented; root large, very hard. Aug.

MENTHA L.

M. spicata L. SPEARMINT. Common along brooks, and in wet grounds. (*M viridis* L.)

M. piperita L. PEPPERMINT. Common along streams, and the borders of swamps; occasionally in dry ground.

M. Canadensis L. WILD MINT. Common in low grounds.

M. citrata Ehrh. BERGAMOT, MINT. Common in waste places. *Graves*. Occasional, *Fenno*. Common in cultivation.

LYCOPUS L.

L. Virginicus L. WATER HOREHOUND. BUGLE-WEED. Common in wet places. Spreads by long stolons.

L. sinuatus Ell. BUGLE-WEED. Common. Found with the preceding.

CUNILA L.

C origanoides (L.) Britton. DITTANY. Very rare. Found on a dry hill on the east bank of the Susquehanna river, nearly opposite the village of Ulster, 1862, *Graves*. Dry hill opposite village of Sayre; plentiful, *Barbour*. These stations are remarkable for being the farthest north from which the plant has been reported. (*C. Mariana* L)

KŒLLIA MŒNCH.

K. Virginiana (L.) MacM. LANCE-LEAVED MOUNTAIN MINT. WILD BASIL. Tolerably common in the eastern part of our range; rare in the western. Not reported from Tioga county. (*Pycnanthemum lanceolatum* Pursh.)

K flexuosa (Walt.) MacM. NARROW-LEAVED MOUNTAIN MINT. Rare. Fields near Binghamton, *Clute*. The only station. Leaves narrower and heads less downy than in the preceding. Bracts lanceolate; calyx-teeth sharp-pointed. (*Pycnanthemum linifolium* Pursh.)

K. incana (L.) Kuntze. MOUNTAIN MINT. Tolerably common from Broome county west. Hills near Oxford; rare, *Coville*. Not reported elsewhere. Found in upland pastures and on dryish slopes. Leaves ovate-oblong, downy beneath, the upper ones hoary on both sides. The whole plant has a pungent mint-like odor. July. Aug. (*Pycnanthemum incanum* Michx.)

ORIGANUM L.

O. vulgare L. WILD MARJORAM. Rare. Ely Hill, *Millspaugh*. Pope's ravine, *Clute*. Reported from Broome county only. Found in dry fields. Aug.

CLINOPODIUM L.

C. vulgare L. BASIL. CALAMINT. Common in dry woods, on banks and in hedges. (*Calamintha Clinopodium* Benth.)

HEDEOMA PERS.

H. pulegioides (L.) Pers. AMERICAN PENNYROYAL. Abundant on dry, warm hills throughout our range.

MONARDA L.

M. didyma L. BEE-BALM. OSWEGO TEA. SCARLET BALM. Common along streams, borders of swamps and in low thickets. One of our most showy wild-flowers; easily cultivated. July. Aug.

M. fistulosa L. WILD BERGAMOT. HORSE-BALM. Infrequent. Found in fields, along roadsides and on banks. Quite variable. The variety, *mollis* Benth., is reported by *Graves* and *Fenno*.

BLEPHILIA RAF.

B. ciliata (L.) Raf.; Benth. Ely Hill, Binghamton, *Millspaugh*. The only station.

VLECKIA RAF.

V. scrophulariæfolia (Willd.) Raf. HEDGE HYSSOP. GIANT HYSSOP. Rare. River bank near Port Dickinson, *Clute*. Barton, *Fenno.* Near Roericke's glen and at Cobble Hill, Chemung county, *Lucy.* The only stations. (*Lophanthus scrophulariæfolius* Benth.)

NEPETA L.

N. Cataria L. CATNIP. CATMINT. Abundant about old buildings, in fields and all waste places.

GLECOMA L.

G. hederacea L. GILL-OVER-THE-GROUND. GROUND IVY. Plentiful in door-yards and waste grounds. (*Nepeta Glechoma* Benth.)

SCUTELLARIA L.

S. lateriflora L. MAD-DOG SKULL-CAP. Frequent in swamps and on the banks of streams. Flowers blue, sometimes white, in axillary, one-sided racemes. Aug.

S. galericulata L. COMMON SKULL-CAP. Plentiful in the same localities as the preceding. Flowers much larger, solitary in the axils of the upper leaves. July, Aug.

PRUNELLA L.

P. vulgaris L. BLUE CURLS. SELF HEAL. Abundant in fields, pastures and roadsides. Flowers blue. Forms with pink and white flowers sometimes occur. (*Brunella vulgaris* L.)

LEONURUS L.

L. Cardiaca L. MOTHERWORT. Common about old buildings, along roadsides and in other waste grounds.

LAMIUM L.

L. maculatum L. DEAD NETTLE. HENBIT. Not common. Sparingly naturalized in waste places. Not reported from Broome and Susquehanna counties.

L. amplexicaule L. In cultivated grounds; not rare, *Lucy.* Barton, *Fenno.* The only stations.

GALEOPSIS L.

G. Tetrahit L. COMMON HEMP-NETTLE. Rather common in waste places.

G. Ladanum L. RED HEMP-NETTLE. Very rare. Pond Brook near a deserted house. The only station.

STACHYS L.

S. aspera Michx. HEDGE NETTLE. Not common. Found in low grounds. Not reported from the Chenango valley.

PLANTAGINACEÆ.

PLANTAGO L.

P. major L. COMMON PLANTAIN. Common in all waste places, about old buildings, etc. Has long been confused with the following species, from which it is distinguished by its shorter, thicker spikes, ovoid capsules circumscissile near the middle, scarious sepals, and leaves of duller green.

P. Rugelii Decaisne. COMMON PLANTAIN. Found in the same situations as the preceding, which it closely resembles. An indigenous species, probably the more common. Distinguished by longer spikes capsules more cylindrical, circumscissile below the middle, and bright green leaves with purplish petioles.

P. lanceolata L. NARROW-LEAVED PLANTAIN. ENGLISH PLANTAIN. RIB-GRASS. Common in all waste grounds.

P. Virginica L. Plentiful one mile south of Barton, *Fenno*. The only station.

ILLECEBRACEÆ.

ANYCHIA MICHX.

A. dichotoma Michx. FORKED CHICKWEED. Sullivan Hill; frequent, *Lucy*. Mt. Prospect; abundant, *Clute*. Not noted elsewhere.

AMARANTHACEÆ.

AMARANTHUS L.

A. hybridus paniculatus (L.) Uline & Bray. RED AMARANTH. Not common. Found in fields and gardens. (*A. paniculatus* L.)

A. retroflexus L. GREEN AMARANTH. A common weed in gardens and waste places.

A. albus L. WHITE AMARANTH. TUMBLE-WEED. PIG-WEED. Common in the same situations as the preceding. Branches from near the base, forming a globular top which in winter breaks loose from its root and is rolled to considerable distances by the wind.

CHENOPODIACEÆ.

CHENOPODIUM L.

C. album L. PIG-WEED. LAMB'S-QUARTERS. RED-ROOT. Abundant in all waste grounds. Frequently used as a pot-herb.

C. hybridum L. MAPLE-LEAVED GOOSEFOOT. Much rarer than the preceding. Found in the same places.

C. capitatum (L.) S. Wats. STRAWBERRY BLITE. Rare. Oakland, *Graves*. The only station.

C. Botrys L. JERUSALEM OAK. Rare. Oakland, *Graves*. Apalachin, *Fenno*. Elmira, *Lucy*. Binghamton, *Millspaugh*. The bruised foliage has a strong odor like turpentine.

ATRIPLEX L.

A. hastata L. Rare. Waste places along the streets of Oxford, *Coville*. (*A. patulum*, var. *hastatum* Gray.)

PHYTOLACCACEÆ.

PHYTOLACCA L.

P. decandra L. POKE-WEED. PIGEON-BERRY. INK-BERRY. GARGET. Found sparingly throughout in waste grounds. A familiar weed. Ink is often made from the berries. The root is reputed to be poisonous, but the young shoots are occasionally used as a pot-herb.

POLYGONACEÆ.

RUMEX L.

R. Patientia L. PATIENCE DOCK. Frequent in meadows and about old barns, *Coville*.

R. Britannica L. GREAT WATER-DOCK. Susquehanna, plentiful, *Graves*. Elmira, infrequent, *Lucy*.

R. verticillatus L. SWAMP DOCK. WATER DOCK. Common in swamps and along ditches and streams.

R. crispus L. CURLED DOCK. YELLOW DOCK. Common in waste places everywhere.

R. obtusifolius L. BITTER DOCK. Not uncommon in gardens and fields.

R. sanguineus L. RED DOCK. Rather rare. Noyes Island, *Millspaugh*. Apalachin, *Fenno*. Sayre, *Barbour*.

R. Acetosella L. FIELD SORREL. HORSE SORREL. SHEEP SORREL. Common everywhere. Most abundant in rather dry sterile soil. Often overruns large areas and is difficult to exterminate. The whole plant has an acid taste.

POLYGONUM L.

P. aviculare L. BIRD'S KNOT-GRASS. LOW KNOT-WEED. Abundant in old dooryards, along paths, and in waste grounds. Well known.

P. erectum L. TALL KNOT WEED. Less common than the preceding, and found in the same places. Whole plant larger and more erect.

P. Pennsylvanicum L. SMART-WEED. Common in wet grounds.

P. amphibium L. WATER SMART-WEED. Not uncommon in wet places. Rare, *Lucy*.

P. emersum (Michx.) Britton. Chemung river at Harrington's Ford, *Lucy*. Binghamton, *Clute*. The only stations. (*P. Muhlenbergii* Watson.)

P. orientale L. PRINCE'S FEATHER. An occasional escape from cultivation, half naturalized along streets and in waste grounds.

P. Persicaria L. LADY'S THUMB. HEART WEED. Very common in all waste places. Leaves marked with a dark triangular spot.

P. Hydropiper L COMMON SMART-WEED. WATER PEPPER. Very common in moist or wet soil.

P. Virginianum L. SLENDER POLYGONUM. Not uncommon in woods and thickets. Flowers in a slender terminal raceme; styles persistent, bent downward. Aug.

P. arifolium L. HALBERD-LEAVED TEAR-THUMB. Plentiful except in the Chemung valley where it is rare. Found in low grounds. Angles of the stem with reflexed prickles.

P. sagittatum L. ARROW-LEAVED TEAR-THUMB. Abundant in moist places. Well known.

P. Convolvulus L. BLACK BIND-WEED. Common in waste grounds, climbing by twining stems.

P. cilinode Michx. FRINGED BIND WEED. Rare, *Graves; Coville*. Sheaths fringed with reflexed hairs.

P. scandens L. CLIMBING BUCK-WHEAT. Common along roadsides and in thickets.

FAGOPYRUM GÆRTN.

F. Fagopyrum (L.) Karst. BUCKWHEAT. Occasionally persists as a weed in old fields. (*F. esculentum* Mœnch.)

ARISTOLOCHIACEÆ.

ASARUM L.

A. Canadense L. WILD GINGER. COLT'S-FOOT. SNAKE-ROOT. Common throughout our range, in rich woods, especially on hillsides. Leaves usually two, kidney-shaped; flowers single, brownish-purple, close to the earth. The rootstock has a pungent, aromatic taste and is said to be a cure for measles and whooping cough. May.

LAURACEÆ.

SASSAFRAS NEES & EBERM.

S. Sassafras (L.) Karst. SASSAFRAS. Common. Found on hillsides in woods and thickets, most frequently in company with the oak. Flowers yellow, preceding the leaves; drupes blue; young or vigorous leaves ovate, entire; others with a lobe on one or both sides. The whole plant has a sweetish, aromatic flavor, which is strongest in the root. (*S. officinale* Nees.)

BENZOIN Fabric.

B. Benzoin (L.) Coulter. SPICE-WOOD. BENJAMIN-BUSH. WILD ALLSPICE. FEVER-BUSH. Common in damp woods and thickets. Flowers yellow, before the leaves; drupes red. The bark has a spicy taste that has been likened to allspice. (*Lindera Benzoin* Blume.)

THYMELÆACEÆ.

DIRCA L.

D. palustris L. LEATHERWOOD. MOOSEWOOD. Tolerably common in damp woods. Flowers yellow, two or more in a cluster, preceding the leaves. Wood soft and very brittle; bark exceedingly tough, often used for thongs. Apr.

DAPHNE L.

D. Mezereum L. MEZEREUM. "A single shrub on the east shore of Cayuta L., 1885. (*F. V. Coville.*)"—*Cayuga Flora*. Probably an escape.

LORANTHACEÆ.

RAZOUMOFSKYA Hoff.

R. pusilla (Peck.) Kuntze. Frequent on *Picea* in the spruce swamps in Chenango county, *Coville*. (*Arceuthobium pusillum* Peck.)

SANTALACEÆ.

COMANDRA Nutt.

C. umbellata (L.) Nutt. BASTARD TOAD-FLAX. Rather common in dryish thickets and open woods. Roots parasitic on the roots of shrubs and trees. Flowers small in corymbose clusters, white.

EUPHORBIACEÆ.

EUPHORBIA L.

E. maculata L. SPOTTED SPURGE. MILK PURSLANE. Abundant, especially in dry, gravelly soil. Plant flat on the ground; leaves marked with a spot of brownish-red.

E. nutans Lag. PRESL'S SPURGE. NODDING SPURGE. Less common than the preceding, which it somewhat resembles. Found in the same places. Stems ascending. (*E. Preslii* Guss.)

E. corollata L. FLOWERING SPURGE. Spanish Hill, *Millspaugh; Graves.* Valley of the Chemung river, *Lucy.*

E. Helioscopia L. SUN SPURGE. WARTWEED Very rare. Norwich, *Millspaugh.* Fields at Unadilla Forks, *Brown.* The only stations.

E. Cyparissias L. CYPRESS SPURGE. TREE MOSS. GRAVE-YARD-WEED. Common along roadsides and in fields, as an escape from old gardens and cemeteries. Flowers yellow. A well-known plant.

E. Nicæensis All. Along road near Gibson, Pa., *Graves.* Vestal, N. Y., *Millspaugh.* Apalachin, Campville, Barton, Athens, *Fenno.* At all but the first station the plant is very abundant and appears to be spreading. Has been naturalized for many years. First noted by Dr. C. F. Millspaugh. Not reported from North America outside of the Susquehanna valley. July, Aug.

E. Peplus L. Not rare in the city of Binghamton, *Clute.* Occasional in gardens, Oxford, *Coville.* Elsewhere not reported.

ACALYPHA L.

A. Virginica L. THREE-SEEDED MERCURY. An abundant weed in fields throughout our range. Well-known by sight, at least

ULMACEÆ.

ULMUS L.

U. pubescens Walt. SLIPPERY ELM. RED ELM. Tolerably common in woods and thickets. A small tree with mucilaginous inner bark that is valued as a medicine. (*U. fulva* Michx.)

U. Americana L. AMERICAN ELM. WHITE ELM. Abundant, especially in low grounds. Valued above all other trees for shade. When grown with room to develop it is the most graceful of our native trees. A specimen twenty-one feet in circumference is growing near Port Dickinson, 1897.

U. racemosa Thomas. CORK ELM. ROCK ELM. Least common of our elms. Remarkable for the large, corky ridges of its bark. Not reported south and east of Broome county.

CELTIS L.

C. occidentalis L. HACK-BERRY. SUGAR-BERRY. NETTLE-TREE. Rare. Banks of the Chenango in the city of Binghamton; numerous trees, *Clute.* Apalachin, scarce; Barton, plentiful, *Fenno.* Elsewhere not reported. A curious tree, "with the bark of an ash, the leaf of an elm and the fruit of a linden." The thin outer rind of the fruit is black, very sweet and edible.

MORACEÆ.

HUMULUS L.

H. Lupulus L. HOP. A well-known plant much cultivated within our territory. It is often found wild, and is apparently native.

——The white and red mulberries (*Morus alba* and *M. rubra*) belong to this family. Neither species seems to occur out of cultivation. The hemp (*Cannabis sativa*), occasionally found in waste ground, also belongs here. It apparently does not persist.

URTICACEÆ.

URTICA L.

U. gracilis Ait. COMMON NETTLE. Tolerably common in rich, moist soil, along fence-rows, about buildings, etc.

URTICASTRUM FABRIC.

U. divaricatum (L.) Kuntze. WOOD NETTLE. Very common in moist, shaded places. (*Laportea Canadensis* Gaud.)

ADICEA RAF.

A. pumila (L.) Raf. RICHWEED. CLEARWEED. Common in damp shade. (*Pilea pumila* Gray.)

BŒHMERIA JACQ.

B. cyclindrica (L.) Willd. FALSE NETTLE. Common in wet places. Flowers minute, clustered, in long axillary spikes which are usually leaf-bearing at the summit.

PLATANACEÆ.

PLATANUS L.

P. occidentalis L. BUTTONWOOD. BUTTONBALL. SYCAMORE. PLANE-TREE. A common and well known tree in river bottoms. Remarkable for shedding its bark in thin scales with its leaves, leaving the surface beneath white or gray-green. Height and diameter of trunk both considered, this is our largest tree.

JUGLANDACEÆ.

JUGLANS L.

J. cinerea L. BUTTERNUT. WHITE WALNUT. Very common, especially along streams. A medium-sized tree with rather smooth, grayish bark. Wood brown, rather hard, much used in cabinet work. The nuts are gathered in considerable quantities.

J. nigra L. BLACK WALNUT. Not common. Most plentiful in the western part of our range. Not reported north and east of Broome county. This seems near the northern limits of the tree's range. A large tree with rough bark and a longer trunk than the preceding. Wood hard, dark, and valued above that of any other of our forest trees for cabinet work.

HICORIA RAF.

H. ovata (Mill.) Britton. SHELLBARK HICKORY. SHAGBARK. Abundant in fields and woods in dryish soil. The nuts are gathered in large quantities for the market. Wood hard, light in color, and in considerable demand in work where strength and elasticity are required. (*Carya alba* Nutt.)

H. sulcata (Willd.) Britton. KING-NUT. BIG SHELLBARK. Rare. Near Waverly, *Millspaugh; Clute*. Newark Valley, *Barbour*. Resembles the preceding. Nut often twice as large. (*Carya sulcata* Nutt.)

H. alba (L.) Britton. MOCKER-NUT. WHITE-HEART HICKORY. Not very common. Not reported from the Chenango valley. Bark rough, but not shaggy. Nut sweet, shell very thick. (*Carya tomentosa* Nutt.)

H. glabra (Mill.) Britton. PIG-NUT. BROOM HICKORY. Common. Nut thick-shelled, the kernel at first sweet then bitter. (*Carya porcina* Nutt.)

H. minima (Marsh.) Britton. BITTER-NUT. SWAMP HICKORY. Less common than the preceding. Nut thin-shelled, kernel bitter. (*Carya amara* Nutt.)

MYRICACEÆ.

COMPTONIA BANKS.

C. peregrina (L.) Coulter. SWEET FERN. Very common in sterile soil, often forming low thickets, especially on hillsides, throughout our range. A well-known low shrub with fern-like, aromatic foliage, in some repute as a medicine. (*Myrica asplenifolia* Endl.)

BETULACEÆ.

BETULA L.

B. lenta L. BLACK BIRCH. SWEET BIRCH. CHERRY BIRCH. Tolerably common in rich soil. Bark with an aromatic odor and sweet, spicy taste. By distillation it yields an oil that is much used as a substitute for oil of wintergreen. Wood hard, dark red, used in cabinet work.

B. lutea Michx. f. YELLOW BIRCH. GRAY BIRCH. Very common in rich, moist woodlands. Bark yellowish, detaching in thin strips that curl up and give the trunk a shaggy appearance. Much less aromatic than the preceding. In spring this species yields sap very freely. Wood soft.

B. populifolia Marsh. WHITE BIRCH. GRAY BIRCH. OLD FIELD BIRCH. Very common, especially in moist soil, in the southern part of our range. Not found in the Chenango and Chemung valleys, Leaves triangular, long pointed. A small tree with chalky-white bark, not aromatic, which does not detach readily in layers. Wood very soft and nearly useless.

B. papyrifera Marsh. PAPER BIRCH. WHITE BIRCH. CANOE BIRCH. Frequent on the northern slopes of hills in the Chemung valley, *Lucy*. Common at Goodrich Lake, *Hoy*. A large clump of trees on the northern slope of South Mountain, *Millspaugh; Clute.* Bark detaching in thin, papery layers, not aromatic. Leaves ovate, taper-pointed.

ALNUS GÆRTN.

A. rugosa (Ehrh.) Koch. SMOOTH ALDER. TAG ALDER. BLACK ALDER. Common along streams, borders of swamps and in all low grounds. Not reported from the Chemung and Chenango valleys. A tall shrub, well known. (*A. serrulata* Willd.)

A. Incana (L.) Willd. HOARY ALDER. SPECKLED ALDER. Not so common as the preceding except in the northern part of our range. Found in the same places.

CORYLUS L.

C. Americana Walt. COMMON HAZEL-NUT. BONNETED HAZEL-NUT. Tolerably common in damp thickets. Rare in Susquehanna and Chenango counties. The nuts are often gathered and go by the name of filberts.

C. rostrata Ait. BEAKED HAZEL-NUT. Plentiful throughout our range in dryish soil, forming thickets. More abundant than the preceding. Nut at the bottom of a bottle-shaped involucre that is densely clothed with slender prickles.

OSTRYA SCOP.

O. Virginiana (Mill.) Willd. IRON-WOOD. HOP-HORNBEAM. LEVER-WOOD. Tolerably common in rich, moist woods. Leaves like the black birch; bark fine and shreddy; fruit resembling hops. A slender tree with exceedingly hard, fine-grained wood, used in making mallets, levers, fish-poles, etc. (*O. Virginica* Willd.)

CARPINUS L.

C. Caroliniana Walt. HORNBEAM. WATER-BEECH. IRON-WOOD. BLUE BEECH. Abundant along streams and in damp woodlands. Bark dark gray, smooth like the beech; fruit somewhat like the preceding. Wood very hard. This species is remarkable for its ridged trunk, which gives it a very muscular appearance.

FAGACEÆ.

QUERCUS L.

Q. alba L. WHITE OAK. Common throughout in any soil. A well-known tree, valued for its hard, durable wood. Fruit annual.

Q. macrocarpa Michx. MOSSY-CUP OAK. BURR OAK. Occasional. Found usually near water. Cup fringed by the long scales often so densely as to hide the acorn within.

Q. platanoides (Lam.) Sudw. SWAMP WHITE OAK. Lowman's swamp; infrequent, *Lucy*. Mutton-Hill Pond, *Fenno*. (*Q. bicolor* Willd.)

Q. Prinus L. ROCK CHESTNUT OAK. Common in upland woods. Not reported from the Chenango valley. Bark very rough and heavy; leaves somewhat like the chestnut.

Q. Muhlenbergii Engelm. YELLOW OAK. CHESTNUT OAK. Somewhat rare. Sullivan Hill, *Lucy*. Barton, *Fenno*. Elsewhere not reported.

Q. prinoides Willd. DWARF CHESTNUT OAK. Frequent, *Lucy*. Barton; Apalachin; not common, *Fenno*. Occasional, *Clute*. Elsewhere not reported. The smallest of our oaks, often fruiting when but two feet high.

Q. rubra L. RED OAK. Common. A well-known large tree, valued for its wood. Acorns large.

Q. coccinea Wang. SCARLET OAK. Tolerably common. Not reported north, east, or south, of Broome county. A large tree, in general appearance resembling the following. Foliage turning bright scarlet in autumn.

Q. velutina Lam. BLACK OAK. QUERCITRON. Not uncommon. Not reported from Susquehanna county. Bark very dark, inner layers orange-colored. Used in dyeing. (*Q. coccinea*, var. *tinctoria* Gray.)

Q. ilicifolia Wang. SCRUB OAK. BEAR OAK. Frequent, *Lucy*. North of Waverly, *Millspaugh*. Shores of a pond two miles west of Greene village, *Coville*. Not reported elsewhere.

Usually found in sterile soil. A low, scrubby tree, said to have formerly been very common in Broome county, but not found there now.

CASTANEA Adans.

C. dentata (Marsh.) Sudw. CHESTNUT. One of the commonest of trees throughout our range, in dryish soils. The nuts are gathered extensively for the markets. (*C. sativa*, var. *Americana* Wats. & Coult.)

FAGUS L.

F. atropunicea (Marsh.) Sudw. COMMON BEECH. Plentiful throughout; usually scattered through the forests, but occasionally forming woods by itself. (*F. ferruginea* Ait.)

SALICACEÆ.

SALIX L.

S. nigra Marsh. BLACK WILLOW. Common throughout. A well-known small tree along all streams and the borders of swamps.

S. nigra falcata Torr. SCYTHE-LEAVED BLACK WILLOW. Common near Waverly, East Waverly and Chemung, *Millspaugh*. (*S. nigra*, var. *falcata* Torr.)

S. lucida Muhl. SHINING WILLOW. GLOSSY WILLOW. Not uncommon on the banks of streams. An elegant species.

S. fragilis L. BRITTLE WILLOW. CRACK WILLOW. East Waverly; frequent, *Millspaugh*.

S. alba vitellina (L.) Koch. WHITE WILLOW. CRACK WILLOW. GOLDEN OSIER. A familiar tree, with thick trunk and slender golden-yellow twigs. Most frequently in cultivation, but often escaped. (*S. alba*, var. *vitellina* Koch.)

S. Babylonica L. WEEPING WILLOW. Everywhere in cultivation and occasionally found wild, where it has probably sprung up from twigs carried by the wind, since both sexes are not found in cultivation.

S. longifolia Muhl. SAND-BAR WILLOW. RIVER-BANK WILLOW. Infrequent. Bottom lands of the Chemung river, *Lucy; Millspaugh*. Not reported elsewhere; probably overlooked.

S. rostrata Richards. BEAKED WILLOW. Not very common; most plentiful in the eastern part of our range.

S. discolor Muhl. GLAUCOUS WILLOW. PUSSY WILLOW. Very common in wet soils. Aments appearing before the leaves in spring; very noticeable.

S. discolor prinoides (Pursh) Anders. Common on the rocky banks of Cayuta creek, near East Waverly, *Millspaugh*. (*S. discolor*, var. *prinoides* Anders.)

S. humilis Marsh. PRAIRIE WILLOW. Tolerably common. A low shrub found on dry hills.

S. tristis Ait. DWARF GRAY WILLOW. SAGE WILLOW. Not uncommon in uplands. Rare, *Lucy*. A shrub much smaller than the preceding, seldom growing more than two feet high.

S. sericea Marsh. SILKY WILLOW. Not uncommon by creeks and swamps throughout.

S. purpurea L. PURPLE WILLOW. Edge of swamp, City of Elmira, *Lucy*.

S. cordata Muhl. HEART-LEAVED WILLOW. Common in wet places.

——Many of the willows hybridize frequently. Hybrids of *S. alba × fragilis*, *S. cordata × sericea*, and forms that may be referred to *S. lucida* and *S. cordata* are reported from the vicinity of Waverly, by Dr. Millspaugh.

POPULUS L.

P. alba L. WHITE POPLAR. ABELE. Not uncommon throughout in cultivation, and spreading by means of suckers from the roots.

P. tremuloides Michx. AMERICAN ASPEN. Abundant in nearly all soils. A familiar small tree of very rapid growth; one of the first to spring up after a wood has been cut down or the soil burned over. Leaves on long petioles, very tremulous in the wind.

P. grandidentata Michx. LARGE-TOOTHED ASPEN. Common. Found with the preceding, which it much resembles, but never in such numbers. Apparently a somewhat larger tree.

P. balsamifera candicans (Ait.) A. Gray. BALM OF GILEAD. In cultivation throughout our range; occasionally escaped along roadsides and banks of streams. (*P. balsamifera*, var. *candicans* Gray.)

P. dilatata L. LOMBARDY POPLAR. Often cultivated, and frequently escaped along roadsides in the vicinity of dwellings.

CERATOPHYLLACEÆ.

CERATOPHYLLUM L.

C. demersum L. HORNWORT. Oxford; common, *Coville*. Lisle; plentiful, *Clute; Graves*. Apalachin; not rare, *Fenno*. The only stations. Found in ponds and slow streams. Plant submerged; leaves cut into thread-like divisions, rendering it very likely to be mistaken at a glance for various species of *Chara*.

CONIFERÆ.

PINUS L.

P. Strobus L. WHITE PINE. Common and well known.

P. rigida Mill. PITCH PINE. TORCH PINE. Much less common than the preceding. Rare in the Chenango valley. Found on dry hillsides, seldom forming woods by itself as the white pine does. Bark dark brownish-red; branches much less regular than in the white pine.

PICEA LINK.

P. Mariana (Mill.) B. S. P. BLACK SPRUCE. Swamp at the head of Christian Hollow, Bradford county, Pa.; rare, *Lucy*. Swamp near Thompson. Pa,; not common, *Graves*. Near Greene, N. Y.; rare, *Clute*. Brisbin swamp and at various points in the northern part of Chenango county. always in swamps, *Coville*. (*P. nigra* Link.)

TSUGA CARR.

T. Canadensis (L.) Carr. HEMLOCK. Very common. Often forming extensive forests, especially in wet or rocky grounds. A familiar tree.

ABIES Juss.

A. balsamea (L.) Mill. BALSAM FIR. "The Vlai," near Oneonta; scarce, *Hoy*. Swamp near Pharsalia, *Coville*. The only stations. In cultivation and occasionally escaped, *Graves*.

LARIX Adans.

L. laricina (Du Roi) Koch. TAMARACK. LARCH. BLACK LARCH. HACKMATACK. In our range found naturally only in cold swamps, *Coville; Lucy; Clute; Hoy*. Elsewhere cultivated as a shade tree, and probably occasionally escaped. Certain swamps are known as "tamarack swamps" from the prevalence of this tree in them. One of the few conifers that are not evergreen. When grown alone it forms a very handsome tree. (*L. Americana* Michx.)

THUJA L.

T. occidentalis L. ARBOR VITÆ. WHITE CEDAR. Not uncommon at Unadilla Forks, *Brown*. Elsewhere occasional in cultivation.

JUNIPERUS L.

J. Virginiana L. RED CEDAR. SAVIN. Rare as a native in our region. Mountain House Narrows; Chemung Narrows, *Lucy*. Elsewhere plentiful in cultivation.

TAXUS L.

T. minor (Michx.) Britton. GROUND HEMLOCK. AMERICAN YEW. Tolerably common on moist, shaded banks and in ravines. A shrub seldom more than five feet high, closely resembling the common hemlock, but with foliage lacking the hemlock's characteristic odor. (*T. Canadensis* Willd.)

HYDROCHARITACEÆ.

UDORA Nutt.

U. Canadensis (Michx.) Nutt. DITCH MOSS. WATER SNAKEWEED. Common in ponds and slow streams. A curious moss-like plant, often forming thick mats in shallow water. (*Elodea Canadensis* Michx.)

VALLISNERIA L.

V. spiralis L. EEL-GRASS. TAPE-GRASS. Common in all the larger streams throughout our range. Whole plant submerged. Leaves narrow, often several feet long. Remarkable from the fact that at the time of flowering the pistillate flowers are raised to the surface of the water on slender scapes, and the stemless staminate flowers break loose from the parent plant, and rise to the surface in order to pollinate them. By the coiling of the scape the fruit is drawn under water to ripen.

ORCHIDACEÆ.

ACHROANTHES RAF.

A. monophylla (L.) Greene. ADDER'S MOUTH. Rare. Swamps, occasional, *Brown*. Oxford; infrequent, *Coville*. The only stations. (*Microstylis monophyllos* Lindl.)

A. unifolia (Michx.) Raf. ADDER'S MOUTH. Very rare. Open woods west of Clarke's stone quarry, Oxford, *Coville*. (*Microstylis ophioglossoides* Nutt.)

LEPTORCHIS THOUARS.

L. Lœselii (L.) MacM. FEN ORCHIS. TWAYBLADE. Rare. Edge of road, upper side of Wellsburg Narrows, *Lucy*. Western side Warn's Pond, Oxford, *Coville*. South Mountain bog, *Millspaugh*. Swamp near Barton, *Fenno*. The only stations. (*Liparis Læselii* Richard.)

CORALLORHIZA R. BR.

C. Corallorhiza (L.) Karst. CORAL-ROOT. Near Oxford; rare, *Coville*. Susquehanna, occasional, *Graves*. (*C. innata* R. Br.)

C. multiflora Nutt. LARGE CORAL-ROOT. Not uncommon in dry, rich woods. Plant leafless, brownish or yellowish in color; lip of the flower three-lobed; spur prominent.

C. odontorhiza (Willd.) Nutt. SMALLER CORAL-ROOT. CRAWLEY-ROOT. Common in the eastern part of our range. Not reported west of Broome county. Similar to the preceding. Plant smaller; lip of flower white, spotted with purple, two-toothed at base. Considered medicinal. July.

LISTERA R. Br.

L. cordata (L.) R. Br. TWAYBLADE. Peat bog, one mile south of Ludlow Pond, town of Smithville, *Coville*. Unadilla Forks; rare, *Brown*.

GYROSTACHYS Pers.

G. cernua (L.) Kuntze. NODDING LADIES' TRESSES. Common in damp, open places. Flowers in spikes, cream-white, sweet scented. Aug., Sept. (*Spiranthes cernua* Richard.)

G. gracilis (Bigel.) Kuntze. SLENDER LADIES' TRESSES. Less common than the preceding. Found on hillsides and in dry, open woods. Flowers white, fragrant, in a twisted spike. (*Spiranthes gracilis* Bigelow.)

PERAMIUM Salisb.

P. repens (L.) Salisb. SMALL RATTLESNAKE PLANTAIN. Somewhat rare. Found in rich, shady woods. Flowers in a loose, one-sided spike. (*Goodyera repens* R. Br.)

P. pubescens (Willd.) C. C. Curtis. COMMON RATTLESNAKE PLANTAIN. Common in rich woods, usually under evergreens. Plants commonly forming little colonies. Leaves veined with white; flowers not in a one-sided spike. Aug. (*Goodyera pubescens* R. Br.)

LIMODORUM L.

L. tuberosum L. GRASS PINK. CALOPOGON. Somewhat rare. Not uncommon in suitable places in the eastern part of our range. Not noted from Broome county west, except in bogs north of Barton (*Fenno*). Reported from Mutton-Hill pond, by Millspaugh in 1885. Does not occur there now. Found in sphagnum bogs. Stem from a small, solid bulb; leaves grass-like; flowers large, three or more, pink-purple. A beautiful orchid. June, July. (*Calopogon pulchellus* R. Br.)

POGONIA Juss.

P. ophioglossoides (L.) Ker. ROSE POGONIA. More common than the preceding with which it is nearly always associated. Confined to the eastern part of our range. Bog near Jarvis street, Binghamton—the only station in Broome county, *Clute*.

Mutton-Hill pond, *Fenno;* *Lucy;*—the station farthest west in our range. Plant often a foot high, with a single ovate leaf half way up the stem and a large, rose-colored, fragrant flower at top, subtended by a bract. June, July.

P. verticillata (Willd.) Nutt. WHORLED POGONIA. Very rare. Glenwood ravine—the only station, *Clute.* Leaves usually five, in a whorl near the top of the stem; flowers inconspicuous, with very long, purplish sepals and straw-colored petals. In aspect so much like *Medeola* that it is probably often mistaken for that species. June.

ORCHIS L.

O. spectabilis L. SHOWY ORCHIS. SPRING ORCHIS. Somewhat rare. "The Vlai," near Oneonta, *Hoy.* Unadilla Forks, *Brown.* Bear Swamp and elsewhere, *Graves.* Pond Brook, *Clute.* South Mountain, *Millspaugh.* Hendy creek, and the "pickaway" on the Chemung, *Lucy.* Tioga Centre, *Fenno.* Oxford, *Coville.* The only stations. West of Broome county there are but three stations for the plant, within our range. Leaves two or three, oblong, near the earth; spike few flowered; flowers purple and white. May.

HABENARIA WILLD.

H clavellata (Michx.) Spreng. THREE-TOOTHED HABENARIA. GREEN WOOD ORCHIS. Swamps near Susquehanna; not rare, *Graves.* Pond Brook; common, *Clute.* Oxford, scarce, *Coville.* Barton, *Fenno.* Elsewhere not reported. Flowers whitish, inconspicuous. July. (*H. tridentata* Hook.)

H. flava (L.) A. Gray. GREEN ORCHIS. Very rare. Oakland, 1894, *Graves.* (*H. virescens* Spreng.)

H. bracteata (Willd.) R. Br. LONG-BRACTED ORCHIS. Very rare. Near Elmira, 1893, *Lucy.*

H. hyperborea (L.) R. Br. LEAFY GREEN ORCHIS. In bogs. Unadilla Forks; occasional, *Brown.* Oxford, occasional, *Coville.* The only stations.

H. Hookeriana A. Gray. HOOKER'S ORCHIS. Frequent in the eastern part of our range. Not reported west of Broome county.

Found in rich, damp woods. In aspect much like the following species, for which it is probably often mistaken. (*H. Hookeri* Torr.)

H. orbiculata (Pursh.) Torr. ROUND-LEAVED ORCHIS. BEAR'S EARS. GALL-OF-THE-EARTH. Common in the eastern part of our range. West of Broome county reported from Cinnamon Lake only (*Lucy*). Found in rich woods, usually in the shade of evergreens. Leaves two, very large, roundish, flat on the ground; flowers greenish-white. June.

H. blephariglottis (Willd.) Torr. WHITE-FRINGED ORCHIS. Peat bog, a mile south of Ludlow swamp, in the town of Smithville, *Coville*. Mutton-Hill Pond, *Millspaugh*. The only stations.

H. lacera (Michx.) R. Br. RAGGED-FRINGED ORCHIS. Bear swamp; Ararat, Pa., *Graves*. Oxford, *Coville*. The only stations.

H. psycodes (L.) A. Gray. PURPLE-FRINGED ORCHIS. Very common in the eastern part of our range, less so in the western. Found in wet, open woods, swamps and other low grounds. Stems often two feet high, leafy; spike large; flowers medium-sized, rose-purple; lip fringed. A very handsome species. June, July.

H. grandiflora (Bigel.) Torr. LARGE PURPLE-FRINGED ORCHIS. Rather rare and unequally distributed. Franklin; frequent, *Hoy*. Susquehanna; not common, *Graves*. Woods near Great-singer's Corners; marshy meadow near Horseheads; rare, *Lucy*. (*H fimbriata* R. Br.)

CYPRIPEDIUM L.

C. parviflorum Salisb. SMALL YELLOW LADY'S SLIPPER. Common, especially in upland woods. Not noted in the upper Chenango valley, *Coville*. Stem leafy; flowers about an inch long, bright yellow, fragrant. With us this plant is *not* partial to low grounds. May.

C. hirsutum Mill. LARGE YELLOW LADY'S SLIPPER. Less common than the preceding. Found growing with it, and also in low grounds. Flowers larger, yellow, scentless. In aspect the plant is much like *C. parviflorum*. May. (*C. pubescens* Willd.)

C. reginæ Walt. SHOWY LADY'S SLIPPER. Unadilla Forks; local but abundant in some swamps, *Brown*. Brisbin swamp; abundant, *Coville*. Swamp near Corning, 1864; very scarce, *Lucy*. South Mountain bog; a single specimen gathered in 1883, *Millspaugh*. The only stations. Stem tall, leafy, bearing at top one or more large, pure white flowers, blotched with purple in front. A superb species. May. (*C. spectabile* Swartz.)

C. acaule Ait. PINK LADY'S SLIPPER. MOCCASIN FLOWER. Frequent, especially in low, wet, coniferous woods. Plant stemless; flowers solitary, pink, rarely white. May.

———All our Cypripediums thrive well in cultivation if given proper soil, shade and moisture.

IRIDACEÆ.

IRIS L.

I. versicolor L. BLUE FLAG. Abundant on the borders of swamps, along streams and in all low grounds.

I. Pseudacorus L. YELLOW FLAG. Along Cayuta creek from a point above Waverly to its mouth, and from thence along the west bank of the Susquehanna to Towanda; an escape from cultivation apparently naturalized, *Lucy; Millspaugh*.

SISYRINCHIUM L.

S. Bermudiana L. BLUE EYED GRASS. Common in fields and meadows. Plant grass like; scapes nearly a foot high, bearing the deep blue or rarely white flowers. June, July. (*S. angustifolium* Mill.)

AMARYLLIDACEÆ.

HYPOXIS L.

H. hirsuta (L.) Coville. STAR-GRASS. Common in the valleys of the Susquehanna and Chemung, but not noted in the Chenango valley. Leaves grass-like; flowers bright yellow. (*H. erecta* L.)

DIOSCOREACEÆ.

DIOSCOREA L.

D. villosa L. WILD YAM. Abundant, *Clute*. Frequent, *Fenno*. Infrequent, *Lucy*. East Waverly, *Millspaugh*. Waverly, *Barbour*. Elsewhere not noted. Found on the river banks twining about the bushes. Leaves heart-shaped; flowers greenish yellow, inconspicuous.

SMILACEÆ.

SMILAX L.

S. herbacea L. CARRION FLOWER. Common in thickets and along fence-rows. Stem tall; leaves ovate or roundish; flowers yellowish-green, in umbels. Very graceful and decorative in appearance, but unapproachable because of the offensive odor of the flowers. June.

S. hispida Muhl. GREEN BRIAR. CAT BRIAR. Occasional in moist woods and thickets. Stem green, armed with numerous prickles; leaves ovate, thin, deciduous.

LILIACEÆ.

ALLIUM L.

A. tricoccum Ait. LEEK Somewhat rare. Near Cohocton, N. Y.,—the only station, *Lucy*. Mouth of Apalachin creek; not frequent, *Fenno*. Along the Chenango and Susquehanna; tolerably common, *Clute*. Oxford, *Coville*. Leaves broad and flat, resembling those of *Erythronium* but unspotted, dying before the flowers appear; flowers ten or more on a scape; bulb strong-scented, edible. July.

A. cernuum Roth. NODDING WILD ONION. Reported only from the western part of our range where it is common. Barton, *Fenno*. Valley of the Chemung, *Lucy*. Along the Susquehanna from Athens to Tunkhannock, *Clute*. Found usually on dryish banks. Scape bearing a few-flowered umbel of drooping, rose-colored flowers; bulb tough, scarcely edible. July.

A. Canadense L. WILD ONION. WILD GARLIC. Abundant on moist banks, in meadows and along streams. Umbels with bulblets among the flowers. Bulb small, strong-scented, tender, mild to the taste, edible. Our commonest species of *Allium*. May.

——The Star-of Bethlehem (*Ornithogalum umbellatum*) and the Grape Hyacinth (*Muscari botrioides*) belong here. They are occasionally reported as escapes, but probably do not persist long in the wild state.

HEMEROCALLIS L

H. fulva L. DAY LILY. Common throughout as an escape that has become well naturalized. Found along roadsides, banks of streams, etc. Flowers large, dull red, several on each scape.

——The Lily-of-the-Valley (*Convallaria majalis*) is occasionally found as an escape but apparently does not spread, although persisting for many years on the sites of old dwellings.

POLYGONATUM ADANS.

P. biflorum (Walt.) Ell. SOLOMON'S SEAL. Common in moist, rich woods, thickets and ravines. Rootstock thick, knotted, creeping, marked with scars where the stems of previous years have been. Stem two feet or more high, somewhat drooping; leaves ovate-oblong, nearly sessile; flowers usually in pairs from the axils of the leaves. In early spring the tips of the young shoots are gathered and used like asparagus. May.

P. biflorum commutatum (R. & S.) Morong. GREAT SOLOMON'S SEAL. Common on the banks of the larger streams throughout our range. In appearance much like the preceding, except that it is very much larger. (*P. giganteum* Dietrich.)

ASPARAGUS L.

A. officinalis L. ASPARAGUS. Not uncommon throughout in fields, meadows and waste grounds. An escape from cultivation that is well naturalized.

VAGNERA ADANS.

V. racemosa (L.) Morong. FALSE SOLOMON'S SEAL. WILD SPIKENARD. Common in moist, rich woods and thickets. Plant similar in appearance to *Polygonatum*, but with the stem ending in a racemose panicle of minute, whitish flowers that are followed by aromatic red berries. May. (*Smilicina racemosa* Desf.)

V. trifolia (L.) Morong. THREE-LEAVED SOLOMON'S SEAL. Tamarack swamp in Preston, three miles from Oxford, *Coville*. (*Smilicina trifolia* Desf.)

UNIFOLIUM ADANS.

U. Canadense (Desf.) Greene. TWO-LEAVED SOLOMON'S SEAL. Very common in moist woods and thickets, especially about the bases of trees and old stumps. Leaves two or three, ovate, cordate at base; flowers white, very small, in a simple raceme. May. (*Maianthemum Canadense* Desf.)

STREPTOPUS MICHX.

S. roseus Michx. TWISTED STALK. Tolerably common in damp woods and thickets. Stem branching, in appearance like a *Polygonatum;* flowers rose-purple, on the under side of the stem, one from the axil of each leaf; peduncle somewhat bent near the middle. May.

S. amplexifolius (L.) DC. TWISTED STALK. Frequent, *Graves*. The only station.

DISPORUM SALISB.

D. lanuginosum (Michx.) Britton. HAIRY DISPORUM. Very rare. Oxford, *Coville*. Binghamton, *Clute*. Kirkwood, Steuben county, *Lucy*. The only stations. May.

CLINTONIA RAF.

C. borealis (Ait.) Raf. NORTHERN CLINTONIA. YELLOW CLINTONIA. Common in cool, moist woods. Leaves two or three, oblong, thick; flowers several on a scape, bell-shaped, yellowish; berry blue. May, June.

UVULARIA L.

U. perfoliata L. MEALY BELLWORT. Common in rich, open woods. Stem about a foot high, forking; flowers one or two, pendulous, straw-colored, the inside of the perianth covered with shining orange or yellow grains. Our only fragrant species. May.

U. grandiflora J. E. Smith. LARGE-FLOWERED BELLWORT. Less common than the preceding which it much resembles. Found in rich woods. Rather tall; flowers large, yellow, the long segments of the perianth twisted. May.

U. sessilifolia L. WILD OATS. Plentiful in woodlands and thickets. In general appearance like the other members of the genus. Flowers rather small, pale yellow. May. (*Oakesia sessilifolia* Watson.)

ERYTHRONIUM L.

E. Americanum Ker. ADDER'S TONGUE. DOG-TOOTH VIOLET. YELLOW LILY. FAWN LILY. Abundant in nearly all low thickets on river banks, etc. Leaves two or three, oblong, spotted with brown; flowers single, drooping, brownish outside, yellow inside, the segments revolute in sunshine. Bulb very deep in the earth. This plant does not flower until its bulbs have reached a sufficient depth. To accomplish this, the young bulbs near the surface send out long runners that form new bulbs at their extremities, upon which the central bulb dies. The large flowering bulbs commonly do not send out runners. One of our earliest spring flowers. The foliage disappears before midsummer.

E. albidum Nutt. WHITE ADDER'S TONGUE. Rare. Oxford, *Coville*. Susquehanna, *Graves*. Much like the preceding. Perianth pinkish-white.

LILIUM L.

L. Philadelphicum L. WILD RED LILY. PHILADELPHIA LILY. WOOD LILY. Abundant in woods, thickets and pastures. Stem simple, leafy; flowers one or two, erect, the segments on claws, spreading, orange or red, spotted with purple and brown inside. Our earliest species. June.

L. superbum L. TURK'S-CAP LILY. SUPERB LILY. Least common of our lilies. Apparently restricted to the borders of our larger streams, but there abundant in some places. Stem tall and stout, bearing numerous large, drooping, orange-yellow flowers, spotted inside with brown; segments of the perianth strongly revolute. An elegant species. Aug.

L. Canadense L. Canada Lily. Meadow Lily. Wild Yellow Lily. Very common in meadows and on the borders of streams. Flowers drooping, yellow or orange, spotted with brown; segments not revolute. July.

———All our lilies do well in cultivation, and often reach a greater size than in the wild state.

MEDEOLA L.

M. Virginiana L. Indian Cucumber-root. Common in moist, rich woods. Stems a foot or more high, the sterile with a whorl of oblanceolate leaves at the top, the fertile with a second smaller whorl above the other, from the center of which depend several greenish-yellow flowers with revolute segments; berries dark red. The rootstock is tapering, white, crisp, edible, and has the taste of cucumbers. May, June.

TRILLIUM L.

T. erectum L. Red Trillium. Wake Robin. Birth-root. Common in damp woods and thickets. Leaves three, rhomboidal, sessile, in a whorl at the top of the stem; flowers single, the three outer segments green, the inner dull purple, deflexed beneath the leaves, ill-scented. An albino form of this, with flowers cream-colored, is the so-called variety *alba*. May.

T. grandiflorum (Michx.) Salisb. White Trillium. Wake Robin. White Lily. Common in woods and thickets. Resembles the preceding. Flower large, erect, the inner segments pure white, turning pink with age. Easy of cultivation. May.

T. undulatum Willd. Painted Trillium. Smiling Wake Robin. Tolerably common in wet woodlands, borders of swamps, and knolls in wet pastures. Leaves petiolate; inner segments of the erect flower spreading, wavy-edged, white striped with purple toward the base. May. (*T. erythrocarpum* Michx.)

CHAMÆLIRIUM Willd.

C. luteum (L.) A. Gray. Blazing Star. Devil's Bit. Not common. Not noted in the upper Chenango valley, *Coville*. Found on dryish hillsides, usually in partial shade. Leaves lanceolate; staminate and pistillate flowers white, on separate plants, in long, wand-like, spiked racemes. June. (*C. Carolinianum* Willd.)

VERATRUM L.

V. viride Ait. WHITE HELLEBORE. INDIAN POKE. Common along streams and in other low grounds. Stem tall, very leafy, leaves ovate, strongly veined and plaited; flowers greenish, in a terminal pyramidal panicle. A characteristic plant in all wet places. Root poisonous.

PONTEDERIACEÆ.

PONTEDERIA L.

P. cordata L. PICKEREL-WEED. Rare. Borders of lakes and ponds, *Graves*. Carmalt Lake, *Fenno*. Elsewhere not reported. Found in shallow water. Leaves cordate, long petioled; flowers blue, in a dense spike. A common plant that is rare with us for want of suitable localities in which to grow.

HETERANTHERA R. & P.

H. dubia (Jacq.) MacM. WATER STAR-GRASS. MUD PLANTAIN. Common in shallow water on the borders of our larger streams. Not always submersed. Leaves long, grass-like; flowers small, yellow, with a long, thread-like tube. Aug. (*H. graminea* Vahl.)

———The common spiderwort (*Tradescantia Virginiana* L.) belongs here. It is occasionally found as an escape about old buildings, but does not persist long in a wild state. Common in cultivation.

JUNCACEÆ.

JUNCUS L.

J. effusus L. COMMON RUSH. BULRUSH. BOG RUSH. Abundant in all swamps and wet grounds.

J. marginatus Rostk. GRASS-LEAVED RUSH. Plentiful in moist grounds, *Graves; Fenno*. Elsewhere not reported.

J. tenuis Willd. SLENDER RUSH. Abundant throughout in wet or dry soil. Frequent along paths.

J. bufonius L. TOAD RUSH. Very plentiful about pools that dry up in summer.

J. articulatus L. JOINTED RUSH. Big Butler Lake and elsewhere not rare, *Graves*. The only stations reported.

J. acuminatus Michx. SHARP-POINTED RUSH. Very common in low grounds. Heads often proliferous.

J. nodosus L. KNOTTED RUSH. Frequent at Oxford, *Coville*.

J. Canadensis J. Gay. CANADA RUSH. Frequent in swamps at Preston, McDonough and German, *Coville*.

JUNCOIDES ADANS.

J. pilosum (L.) Kuntze. HAIRY WOOD RUSH. Common throughout in moist woodlands and on shady banks. Plant grass-like; flowers like minute brownish lilies. May. (*Luzula vernalis* DC.)

J. campestre (L.) Kuntze. COMMON WOOD RUSH. Plentiful in open woods in dry or moist soil. Much resembles the preceding, but with flowers in close clusters. (*Luzula campestris* DC.)

TYPHACEÆ.

TYPHA L.

T. latifolia L. CAT-TAIL. Abundant in swamps throughout our range. A well-known plant.

T. angustifolia L. NARROW-LEAVED CAT-TAIL. In ditches along the railway at Oxford, possibly introduced from passing trains, *Coville*. The only station.

SPARGANIACEÆ.

SPARGANIUM L.

S. eurycarpum Engelm. BROAD-FRUITED BUR-REED. Common in marshes and the borders of streams. Leaves somewhat resembling the cat-tail; fruit assembled in globular heads.

S. simplex Huds. SIMPLE-STEMMED BUR-REED. Less common than the preceding. Plant much smaller; inflorescence not branched.

S. androcladum (Engelm.) Morong. BRANCHING BUR-REED. Miller's Pond, Chemung county, *Lucy*. The only station. (*S. simplex*, var. *androcladum* Engelm.)

ARACEÆ.

ARISÆMA Mart.

A. triphyllum (L.) Torr. INDIAN TURNIP. JACK-IN-THE-PULPIT. Very common in moist woodlands and ravines. Leaves usually two, three-parted; scape single, the spathe surrounding and arching over the spadix; corm turnip-shaped, very acrid when green. Medicinal.

A. Dracontium (L.) Schott. GREEN DRAGON. Plentiful along the banks of our larger streams. Rare, *Lucy*. Leaf usually solitary, with five to seven leaflets; spadix attenuate and much longer than the spathe.

PELTANDRA Raf.

P. Virginica (L.) Kunth. GREEN ARROW-ARUM. Cinnamon Lake; plentiful, *Lucy*. The only station. (*P. undulata* Raf.)

CALLA L.

C. palustris L. WILD CALLA. WATER ARUM. Common in swamps and on boggy shores. Not reported from the Chemung valley. Rhizome extensively creeping; flowers much resembling those of the cultivated calla lily, but smaller; spathe flat and open; occasionally blooms in autumn. An interesting plant, easily cultivated. May, June.

SPATHYEMA Raf.

S. fœtida (L.) Raf. SKUNK CABBAGE. Abundant on the margin of swamps and in all low grounds. Leaves very large, veiny; flowers preceding the leaves; spathe shell-shaped, surrounding the globular spadix. A characteristic plant in wet places. The bruised foliage has an odor like that of the skunk. One of the earliest flowers to bloom. March. (*Symplocarpus fœtidus* Salisb.)

ORONTIUM L.

O. aquaticum L. GOLDEN CLUB. Rare. Brisbin Pond; Jam Pond, *Coville*. Bog near Jarvis street, Binghamton; Thompson's marsh; plentiful at the latter station, *Clute*. Elsewhere not reported. This plant is abundant in certain lakes just south of our range, where it is reported to have been planted by Indians. It may possibly occur at a few other stations within our limits. Found always in very wet soil. often in standing water. Leaves oblong, on long petioles; spadix near the tip of a long, slender scape. A curious plant, well named from its inflorescence—a golden club.

ACORUS L.

A. Calamus L. SWEET FLAG. CALAMUS-ROOT. Tolerably common on the borders of swamps and along streams. Leaves resembling those of *Typha;* scapes similar to the leaves, the spadix borne on one edge near the top; rootstock thick, creeping, valued for its aromatic flavor and warm, pungent taste.

LEMNACEÆ.

SPIRODELA SCHLEID.

S. polyrhiza (L.) Schleid. DUCK WEED. DUCK MEAT. Very abundant in stagnant pools and slow streams throughout our range, floating on the surface of the water. Thallus small, round-obovate, with a cluster of rootlets from the center of the under surface. In autumn the fronds develop minute bulblets, which sink to the bottom to rise and vegetate in spring. One of the smallest of flowering plants, but often found in such abundance as to hide the water in which it grows.

LEMNA L.

L. minor L. SMALL DUCK WEED. Less common than the preceding and found in the same places. Thallus obovate; root solitary.

L. trisulca L. IVY-LEAVED DUCK WEED. STAR DUCK WEED. Brisbin swamp, *Coville*. The only station.

ALISMACEÆ.

ALISMA L.

A. Plantago-aquatica L. WATER PLANTAIN. Common in swamps, along streams and in other low grounds, often in shallow water. Leaves ovate or oblong; panicle a foot or more high, with numerous, small, white or pale pink, three-parted flowers. June-Aug. (*A. plantago* L.)

SAGITTARIA L.

S. latifolia Willd. COMMON ARROW-HEAD. BROAD-LEAVED ARROW-HEAD. Frequent in wet ground or shallow water. A characteristic plant on the borders of our larger streams. Quite variable as to foliage. (*S. variabilis* Engelm.)

S. rigida Pursh. SESSILE-FRUITING ARROW-HEAD. Less common than the preceding; found in the same places. Leaves variable, usually narrow, entire or with slender basal lobes. (*S. heterophylla* Pursh.)

S. graminea Michx. GRASS-LEAVED ARROW-HEAD. Geneganslet Lake; abundant, *Coville*. Coves of the Chenango and Susquehanna; plentiful, *Clute*. Along the Susquehanna in shallow water, *Graves; Fenno*. Not reported elsewhere. Leaves long-petioled, the blades linear to elliptic, often submerged.

JUNCAGINACEÆ.

SCHEUCHZERIA L.

S. palustris L. Very rare Round Pond, McDonough; Brisbin Pond, *Coville*. Pond Brook, *Clute*. "Ararat marsh, *Dudley*." —*Flora of the Lackawanna and Wyoming Valleys*. The only stations.

NAIADACEÆ.

POTAMOGETON L.

P. natans L. COMMON POND WEED. Common in slow streams and pools throughout our range.

P. Nuttallii Cham. & Schlecht. Almost as common as the preceding in ponds and streams. (*P. Pennsylvanicus* Cham.)

P. Spirillus Tuckerm. SPIRAL POND WEED. "Cayuta Lake."—*Cayuga Flora*.

P. lonchites Tuckerm. Susquehanna river, *Graves*. Chemung river, *Lucy*. Chenango river, *Coville*. (*P. fluitans* Roth.)

P. amplifolius Tuckerm. LARGE-LEAVED POND WEED. Goodrich Lake, *Hoy*. Oxford, *Coville*. "Cayuta Lake."—*Cayuga Flora*.

P. heterophyllus Schreb. "Cayuta Lake."—*Cayuga Flora*.

P. perfoliatus L. CLASPING-LEAVED POND WEED. Chenango river, *Coville*. Chemung river, *Lucy*. Susquehanna river, *Graves*.

P. crispus L. Chemung river; not rare, *Lucy*.

P. zosteræfolius Schum. Chenango river, *Coville*. Susquehanna river, *Graves*.

P. pectinatus L. FENNEL-LEAVED POND WEED. Chemung river, *Lucy*. Chenango river, *Coville*. Susquehana river, *Graves*.

P. foliosus Raf. Chenango river at Oxford, *Coville*. (*P. pauciflorus* Pursh.)

——As yet our pond weeds are but little known. Further study of this group will certainly disclose more species within our boundaries, and extend the list of stations for those that have been identified.

ZANNICHELLIA L.

Z. palustris L. HORNED POND WEED. Below the feeder dam in the Chenango river north of Oxford, *Coville*. The only station.

NAIAS L.

N. flexilis (Willd) Rost. & Schmidt. Oxford, *Coville*. Unadilla Forks, *Brown*. The only stations reported.

ERIOCAULACEÆ.

ERIOCAULON L.

E. septangulare With. SEVEN-ANGLED PIPEWORT. Cinnamon Lake, abundant, *Lucy*. The only station.

CYPERACEÆ.

CYPERUS L.

C. diandrus Torr. LOW CYPERUS. Common in low grounds and on the banks of rivers and streams. Annual.

C. rivularis Kunth. SHINING CYPERUS. Along Apalachin Creek with the preceding, *Fenno*. Scales dark chestnut color. (*C. diandrus*, var. *castaneus* Torr.)

C. aristatus Rottb. AWNED CYPERUS. Rare. Harrington's Island, *Lucy*. Shores of the Susquehanna, *Clute*. The only stations. This species is very small and may easily be overlooked.

C. dentatus Torr. TOOTHED CYPERUS. Pond Brook; not common, *Clute*. The only station. Scales reddish-brown, tips spreading, appearing as if toothed.

C. esculentus L. YELLOW NUT-GRASS. Harrington's Island, *Lucy*. Apalachin Creek, *Fenno*. Oxford, *Coville*. Near Susquehanna, *Graves*. Elsewhere not reported. Root usually bearing small tubers. Sometimes becomes troublesome as a weed, but with us the plant is rare.

C. strigosus L. STRAW-COLORED CYPERUS. Common throughout in low grounds. A variable species.

DULICHIUM L. C. RICHARD.

D. arundinaceum (L.) Britton. COMMON DULICHIUM. Plentiful on the borders of swamps and streams. (*D. spathaceum* Pers.)

ELEOCHARIS R. BR.

E. ovata (Roth) Roem. & Schult. OVATE SPIKE-RUSH. Very common throughout our range in wet and muddy places. Well known to sight at least. Bristles of the spike six to eight.

E. palustris (Roth) Roem. & Schult. SPIKE-RUSH. Not common. Butler Lake and borders of ponds, *Graves*. Apalachin, in shallow water, *Fenno*. Bristles, four.

E. acicularis (L.) Roem. & Schult. NEEDLE-SPIKE-RUSH. Very common on the borders of lakes and streams at the very edge of the water. Stems usually less than six inches high, often forming thick mats over considerable areas. Bristles three or four, falling early.

SCIRPUS L.

S. planifolius Muhl. WOOD CLUB-RUSH. Frequent on dry knolls, *Lucy*. Plentiful on dry banks at Apalachin, *Fenno*. Elsewhere not reported; probably overlooked. Bristles four to six, barbed upward. The bristle-tipped scale overtopping the chestnut-brown spikelet.

S. lacustris L. GREAT BULL-RUSH. Common on the borders of lakes and the larger streams, nearly always in shallow water. Stems round and very tall.

S. atrovirens Muhl. DARK-GREEN BULL-RUSH. Common about swamps, ditches and in low grounds generally throughout our range. Our earliest Scirpus.

S. polyphyllus Vahl. LEAFY BULL-RUSH. Not uncommon, *Graves*. Elsewhere not reported. Spikelets yellow-brown, clustered in threes and eights, division of the umbel short. Bristles six, flexuous. Resembles *S. atrovirens*. Verified by Prof. Thomas C. Porter.

S. cyperinus (L.) Kunth. WOOL-GRASS. COTTON-GRASS. Very plentiful in low grounds. (*Eriophorum cyperinum* L.)

S. cyperinus Eriophorum (Michx.) Britton. Probably more common than the type, from which it is distinguished by its long-pedicelled, lateral spikes. (*Eriophorum cyperinum*, var. *laxum* Gray.)

ERIOPHORUM L.

E. Virginicum L. VIRGINIA COTTON-GRASS. Common in bogs and swamps throughout our range, except in the Chemung valley, where it is probably overlooked. Wool rusty, or copper-colored; spikelets nearly sessile.

E. polystachyon L. TALL COTTON-GRASS. Frequent. Found in the same situations as the preceding. Culms a foot or more high; wool white, copious, in tufts like cotton; noticeable at a considerable distance. Dr. Lucy observes that this species does not occur in the Chemung valley.

RHYNCHOSPORA Vahl.

R. alba (L.) Vahl. WHITE BEAK-RUSH. Rather rare. Pond Brook, frequent, *Clute*. Mutton-Hill Pond; bogs north of Barton, *Fenno*. The only stations.

CLADIUM P. Br.

C. mariscoides (Muhl,) Torr. TWIG-RUSH. The only known station for this plant within our limits is at Pond Brook, where it is frequent, *Clute*.

CAREX L.

C. pauciflora Lightf. FEW-FLOWERED SEDGE. Very rare. "Jones' Lake, near Montrose, Pa. (Garber, 1869). The southern limits of this species."—List of the Carices of Pennsylvania, *Porter*.

C. abacta Bailey. YELLOWISH SEDGE. Very rare. A single specimen found on the borders of Beebe's swamp and now in the Herbarium of Lafayette College. This is the only specimen reported from the State of Pennsylvania, *Graves*. (*C. Michauxiana* Bœckl.)

C. intumescens Rudge. BLADDER SEDGE. Abundant in wet pastures and swamps.

C. Asa-Grayi Bailey. GRAY'S SEDGE. Plentiful near Barton, *Fenno*. Specimen verified by Dr. C. H. Peck. (*C. Grayii* Carey.)

C. lupulina Muhl. HOP SEDGE. Common. Found on the borders of swamps throughout our range. Spikes hop like in aspect.

C. utriculata Boott. BOTTLE SEDGE. Tolerably common on the borders of ponds and swamps.

C. monile Tuckm. NECKLACE SEDGE. Somewhat rare. Susquehanna, *Graves*. Franklin, *Hoy*. The only stations.

C. Tuckermani Dewey. TUCKERMAN'S SEDGE. More common than the preceding, which it much resembles. Found throughout our range

C. bullata Schk. BUTTON SEDGE. Rare Wet ditch near Brushville; Butler's Lake, *Graves*. The only stations. Specimen identified by Prof. T. C. Porter.

C. retrorsa Schwein. RETRORSE SEDGE. Not uncommon in swamps and wet meadows.

C. retrorsa Hartii (Dewey) A. Gray. Very rare. Swale near Beebe's swamp, *Graves*. Verified by Prof. T. C. Porter. One of the rarest of Pennsylvania carices.

C. lurida Wahl. SALLOW SEDGE. Plentiful throughout our range in wet and sometimes in dry soil.

C. lurida gracilis (Boott.) Bailey. Found in the same places as the preceding, with which it is often confused.

C. Schweinitzii Dewey. SCHWEINITZ'S SEDGE. Very rare. Found at Beebe's swamp. Never reported from Pennsylvania but once before and then by Schweinitz himself, *Graves*.

C. hystricina Muhl. PORCUPINE SEDGE. Less frequent than *C. lurida*, with which it is often confounded. Found in marshes and low ground.

C. Pseudo-Cyperus L. CYPERUS-LIKE SEDGE. Somewhat rare. Lowe's Pond, town of Big Flats, *Lucy* Butler's Lake; Beebe's Swamp, *Graves*. Pond Brook, *Clute*. Elsewhere not reported.

C. Pseudo-Cyperus Americana Hochst. More common than the type and found in similar places.

C. scabrata Schwein. ROUGH SEDGE. Common in wet shades throughout.

C. filiformis lanuginosa (Michx.) B. S. P. WOOLLY SEDGE. Not common. Botany Swamp. Elmira, *Lucy*. Bear Swamp, *Graves*. Pond Brook, *Clute*. Not noted elsewhere. (*C. filiformis* var. *latifolia* Boeckl.)

C. trichocarpa Muhl. HAIRY-FRUITED SEDGE. Mutton-Hill Pond, *Fenno*. Lisle, N. Y.; not rare, *Graves; Clute*. Not noted elsewhere. The variety *imberbis* is reported from Beebe's Swamp. (*Graves.*)

C. riparia Curtis. RIVER-BANK SEDGE. Jenning's Swamp, New Milford; not plentiful, *Graves*. Binghamton; common in swamps, *Clute*. In marshes, Apalachin; frequent, *Fenno*.

C. stricta Lam. TUSSOCK SEDGE. Common on river banks and elsewhere. *Graves; Fenno*.

C. stricta angustata (Boott) Bailey. Common with the type, *Graves*.

C. stricta decora Bailey. Frequent with the type, *Graves*.

C. torta Boott. TWISTED SEDGE. Most plentiful in the eastern part of our range. Found on the banks of streams.

C. prasina Wahl. DROOPING SEDGE. Not uncommon at Susquehanna, *Graves*. Rare, *Lucy*. Elsewhere not reported.

C. crinita Lam. DROOPING SEDGE. FRINGED SEDGE. Common in swamps and on the banks of streams. A noticeable and elegant species. Achenes curved. The variety *gynandra* is reported by *Lucy* and *Fenno*.

C. crinita minor Boott. Common with the type, *Graves*.

C. limosa L. MUD SEDGE. Pond Brook; not common, *Clute*. The only station.

C. virescens Muhl. DOWNY GREEN SEDGE. Common on dry banks and grassy places, *Graves*. Elsewhere not reported; probably overlooked.

C. virescens costata (Schw.) Dewey. Occasional in dry soil with the type, *Graves*. Quite distinct from the preceding.

C. triceps hirsuta (Willd.) Bailey. HAIRY SEDGE. Barton; rare. *Fenno*. Elmira; not frequent, *Lucy*. Foliage bright green.

C. longirostris Torr. LONG-BEAKED SEDGE. Somewhat rare, found on rich river banks. Noticeable on account of its long beak and long acuminate scale.

C. arctata Boott. DROOPING WOOD SEDGE. Most frequent in the eastern part of our range. Found along roadsides in open woods.

C. debilis Rudgei Bailey. SLENDER-STALKED SEDGE. Found in the same places as the preceding, which it closely resembles, *Graves; Fenno.*

C. gracillima Schwein. GRACEFUL SEDGE. Common throughout in moist woodlands.

C. grisea Wahl. GRAY SEDGE. Frequent throughout in woodlands and thickets, wet or dry. Perigynium plump and beakless.

C. granularis Muhl. MEADOW SEDGE. Rare, *Lucy.* Infrequent, *Fenno.* Not common, *Clute.* Not reported from Susquehanna county.

C. flava L. YELLOW SEDGE. Pond Brook; not uncommon, *Clute.* The only station.

C. pallescens L. PALE SEDGE. Most common in the eastern part of our range. Found in fields, along roadsides and on banks. Easily confused with *C. granularis*, but distinguished from it by the beakless perigynium.

C. laxiflora Lam. LOOSE-FLOWERED SEDGE. Very common in open woods and thickets throughout our range.

———The varieties *patulifolia, blanda* and *varians* of *C. laxiflora* are reported from the same localities as the type. Doubtless the other varieties of this extremely variable species may yet be found.

C. styloflexa Buckley. BENT SEDGE. In moist soil at Barton; infrequent, *Fenno.* (*C. laxiflora*, var. *styloflexa* Boott.)

C. albursina Sheldon. WHITE BEAR SEDGE. Not uncommon in rich shaded soil, especially in ravines. (*C. laxiflora*, var. *latifolia* Boott.)

C. digitalis Willd. SLENDER WOOD SEDGE. Somewhat rare. Found in dryish open woods and thickets. Leaves bright green; perigynium slightly bent.

C. laxiculmis Schwein. SPREADING SEDGE. Cannavan's Glen and elsewhere; not common, *Graves*. Lowman's Swamp woods; rare, *Lucy*. Elsewhere not reported.

C. plantaginea Lam. PLANTAIN-LEAVED SEDGE. Fairly plentiful in dry, shady places, in rich soil. An early species.

C. Pennsylvanica Lam. PENNSYLVANIA SEDGE. Abundant in all dry thickets and on half-shaded knolls. Probably our commonest species and one of the earliest to bloom. A noticeable species in spring, its yellow anthers contrasting strongly with its reddish staminate scales.

C. pedicellata (Dewey) Britton. FIBROUS-ROOTED SEDGE. Tolerably common in dry woods and thickets. (*C. communis* Bailey.)

C. umbellata Schk. UMBEL-LIKE SEDGE. Plentiful along old wood-roads in thin, grassy places, *Graves*. Spikes often hidden in the old leaves, hence unobserved.

C. pubescens Muhl. PUBESCENT SEDGE. Infrequent. Susquehanna, *Graves*. Barton, *Fenno*. Elsewhere not reported.

C. leptalea Wahl. BRISTLE-STALKED SEDGE. Common on the borders of swamps. Not reported from the Chemung valley. (*C. polytrichoides* Muhl.)

C. stipata Muhl. AWL-FRUITED SEDGE. Common throughout on the borders of ditches and swamps. Spikelets terminal in an oblong cluster.

C. vulpinoidea Michx. FOX SEDGE. Common throughout in low grounds.

C. tenella Schk. SOFT-LEAVED SEDGE. Pond Brook. Not uncommon, *Clute*. Cranberry marsh, *Graves*. Elsewhere not reported.

C. rosea Schk. STELLATE SEDGE. Common in moist woods and thickets.

C. rosea radiata Dewey. Similar in appearance, abundance and distribution to the preceding.

C. retroflexa Muhl.   REFLEXED SEDGE.   Occasional in rich upland woods.   (*C. rosea*, var. *retroflexa* Torr.)

C. sparganioides Muhl.   Flentiful in shady places, wet or dry.

C. cephaloidea Dewey.   THIN-LEAVED SEDGE.   Not common. Found on dry hills.   Reported by *Graves* and *Lucy* only.

C. cephalophora Muhl.   OVAL-HEADED SEDGE.   Common on dry knolls and in open woodlands.

C. sterilis Willd.   LITTLE PRICKLY SEDGE.   Not uncommon in swamps and bogs.   The variety *cephalantha* is reported occasional by *Graves*.   (*C. echinata*, var. *microstachys* Bœckl.)

C. canescens L.   SILVERY SEDGE.   Not uncommon in swamps and the borders of ponds.

C. trisperma Dewey.   THREE-FRUITED SEDGE.   Infrequent. Found plentiful in a few small areas.   Susquehanna, *Graves*. Pond Brook, *Clute*.   Barton, *Fenno*.   Elsewhere not reported.

C. Deweyana Schwein.   DEWEY'S SEDGE.   Plentiful in dry, open woods.

C. bromoides Schk.   BROME-LIKE SEDGE.   Common in shaded swamps and wet woodlands.

C. tribuloides Wahl.   BLUNT BROOM SEDGE.   Common in low grounds and swamps.   Spikes usually clustered in a blunt, heavy head.

C. tribuloides Bebbii (Olney) Bailey.   Occasional with the type, *Graves*.   Heads very dense.

C. scoparia Schk.   POINTED BROOM SEDGE.   Very common in wet, open places everywhere.   Distinguished by its tawny, sharp-pointed spikes.

C. straminea Willd.   STRAW SEDGE.   Common in dry fields and on banks throughout our range.   One of our earliest species.

C. straminea festucacea (Willd.) Tuckm.   Beebe's flats; not rare, *Graves*.   (*C. straminea*, var. *brevior* Bailey.)

GRAMINEÆ.

PANICUM L.

P. lineare Krock. SMOOTH CRAB-GRASS. Common and troublesome, *Fenno; Hoy*. (*P. glabrum* Gaud.)

P. sanguinale L. COMMON CRAB-GRASS. FINGER-GRASS. Common in cultivated ground throughout our range.

P. proliferum Lam. On an island in the Susquehanna at Apalachin; rare, *Fenno*.

P. capillare L. OLD WITCH GRASS Common and troublesome in all cultivated grounds

P. elongatum Pursh. AGROSTIS LIKE PANICUM. Tolerably common on wet shores. Not reported from the Chemung valley. (*P. agrostoides* Muhl.)

P. virgatum L. Along the Susquehanna throughout Tioga county, *Fenno*.

P. xanthophysum Gray. SLENDER PANICUM. "Near Painted Post."—*Cayuga Flora*.

P. Walteri Poir. Common in open upland woods and thickets. (*P. latifolium* L.)

P. clandestinum L. HISPID PANICUM. Common in thickets and on river banks. Not reported from the Chemung valley.

P. commutatum Schultes. Apalachin, *Fenno*. The only station.

P. depauperatum Muhl. STARVED PANICUM. Common in dry soils.

P. pubescens Lam. HAIRY PANICUM. Common in dry fields and waste places.

P. dichotomum L. FORKED PANICUM. Very common in dry woods and fields.

P. Crus-galli L. BARN-YARD GRASS. COCKSPUR GRASS. Very common in dry soils throughout our range.

P. Crus-galli hispidum (Muhl.) Torr. Common on river banks and in moist grounds. A larger and more hairy form of the preceding.

P. miliaceum L. MILLET. Occasionally found in waste places throughout.

CHAMÆRAPHIS R. BR.

C. glauca (L.) Kuntze. FOX TAIL. PIGEON GRASS. Very common in dry fields and along roadsides throughout our range. (*Setaria glauca* Beauv.)

C. viridis (L.) Porter. GREEN FOX TAIL. Less common than the preceding and found in the same place. (*Setaria viridis* Beauv.)

———The millet (*C. Italica*) is frequent in cultivation and occasionally persists along roadsides and in old fields for some time.

HOMALOCENCHRUS MIEG.

H. Virginicus (Willd.) Britton. WHITE GRASS. Common throughout in damp, shady places. (*Leersia Virginica* Willd.)

H. oryzoides (L.) Poll. CUT GRASS. FALSE RICE. Frequent in swamps and along streams. (*Leersia oryzoides* Swartz.)

ANDROPOGON L.

A. provincialis Lam. FORKED BEARD-GRASS. BIG BLUE-STEM. Not uncommon on the banks of our larger streams. (*A. furcatus* Muhl.)

A. scoparius Michx. LITTLE BLUE STEM Less common than the preceding. Usually found in rather dry ground.

A. nutans avenaceus (Michx) Hack. INDIAN GRASS. WOOD GRASS. Tolerably common on dry, shady banks, especially along streams. (*Chrysopogon nutans* Benth.)

PHALARIS L.

P. arundinacea L. REED CANARY-GRASS. Infrequent. Found on the borders of lakes and streams. The variety *picta* with white-striped leaves is the familiar ribbon-grass of old gardens.

———The canary-grass (*P. Canariensis*) is not uncommon in waste places. It apparently does not persist long.

ANTHOXANTHUM L.

A. odoratum L. SWEET VERNAL-GRASS. Frequent along roadsides and in fields. Rare, *Lucy*. Very sweet-scented in drying.

ORYZOPSIS Michx.

O. melanocarpa Muhl. BLACK MOUNTAIN RICE. Plentiful on a rocky hillside west of Barton, *Fenno.* Oxford, *Coville.* Elsewhere not reported.

O. asperifolia Michx. MOUNTAIN RICE. Frequent in moist upland woods, *Graves; Fenno; Lucy.*

MILIUM L.

M. effusum L. WILD MILLET. Tolerably common in wet shady places.

MUHLENBERGIA Schreb.

M. sobolifera (Muhl.) Trin. ROCK MUHLENBERGIA. Not rare; in rocky woods, *Graves.*

M. racemosa (Michx.) B. S. P. WILD TIMOTHY. MARSH MUHLENBERGIA. Frequent in wet grounds, *Graves; Fenno.* Elsewhere not reported. (*M. glomerata* Trin.)

M. Mexicana (L.) Trin. MEADOW MUHLENBERGIA. Not uncommon in low grounds throughout our range.

M. sylvatica (Torr.) A. Gray. WOODLAND DROP-SEED. Common on the wooded banks of streams, *Fenno; Graves; Coville.* Elsewhere not reported; probably overlooked.

M. diffusa Schreb. DROP-SEED. NIMBLE-WILL. Common at Apalachin and Barton, *Fenno.* Elsewhere not reported.

BRACHYELYTRUM Beauv.

B. erectum (Schreb.) Beauv. BEARDED SHORT-HUSK. Frequent in moist upland woods throughout our range. (*B. aristatum* Beauv.)

PHLEUM L.

P. pratense L. TIMOTHY. HERD'S GRASS. Abundant in cultivation and naturalized everywhere. The principal meadow grass of our region.

ALOPECURUS L.

A. pratensis L. MEADOW FOXTAIL. West of Apalachin; not common, *Fenno.*

A. geniculatus L. FLOATING FOXTAIL. South of Apalachin in a small stream; infrequent, *Fenno.*

SPOROBOLUS R. Br.

S. vaginæflorus (Torr.) Wood. Not uncommon throughout Tioga county, *Fenno*.

AGROSTIS L.

A. alba L. RED-TOP. WHITE BENT-GRASS. Plentiful in fields. Cultivated for hay. Probably not indigenous.

A. alba vulgaris (With.) Thurb. RED TOP. HERD'S GRASS. Common in low meadows. Very similar to the preceding.

A. perennans (Walt.) Tuckerm. THIN GRASS. Plentiful in shaded places.

A. hiemalis (Walt.) B. S. P. HAIR-GRASS. Not uncommon in shaded places. (*A. scabra* Willd.)

CINNA L.

C. arundinacea L. WOOD REED GRASS. INDIAN REED. Common on the borders of ponds and streams.

C. latifolia (Trev.) Griseb. SLENDER REED-GRASS. Not uncommon. Found with the preceding. (*C. pendula* Trin.)

CALAMAGROSTIS Adans.

C. Canadensis (Michx.) Beauv. BLUE-JOINT GRASS. Not uncommon in swamps, wet woods and along streams.

HOLCUS L.

H. lanatus L. VELVET GRASS. SOFT MEADOW-GRASS. In moist meadows; frequent, *Graves*. Rare, *Fenno*. Elsewhere not reported.

DESCHAMPSIA Beauv.

D. flexuosa (L.) Trin. WAVY HAIR-GRASS. River bank at Apalachin, *Fenno*. Rocky upland woods, near Elmira, *Lucy*. Elsewhere not noted.

AVENA L.

A. striata Michx. WILD OATS. Infrequent. Found on rocky hillsides, *Graves; Fenno; Lucy*. Elsewhere not reported; probably overlooked.

DANTHONIA DC.

D. spicata (L.) Beauv. WIRE-GRASS. Very common, especially in dry, sterile soil. This grass often occupies large patches in meadows, but is disliked by the farmer because its thin, wiry stems yield little hay.

D. compressa Austin. FLATTENED OAT-GRASS. Apalachin in shade, *Fenno*.

KŒLERIA Pers.

K. cristata (L.) Pers. Sullivan Hill; in dry, open woods, *Lucy*. Dry bank at Apalachin; plentiful, *Fenno*.

EATONIA Raf.

E. Pennsylvanica (DC.) A. Gray. Not uncommon in moist thickets.

E. Dudleyi Vasey. Binghamton; rare, *Clute*.

ERAGROSTIS Beauv.

E. hypnoides (Lam.) B. S. P. CREEPING MEADOW-GRASS. Plentiful along Apalachin Creek and west of Apalachin, *Fenno*. (*E. reptans* Nees.)

E. major Host. PUNGENT MEADOW-GRASS. Rare. Roadside, village of Horseheads, *Lucy*. Along the railway at Apalachin and Campville, *Fenno*.

E. pilosa (L.) Beauv. Apalachin in thin, sandy or gravelly soil, *Fenno*.

E. Caroliniana (Spreng.) Scribn. Common at a sand bank at Apalachin, *Fenno*. (*E. Purshii* Schrad.)

E. Frankii Steud. Abundant in an old gravel pit two miles west of Apalachin, *Fenno*.

DACTYLIS L.

D. glomerata L. ORCHARD GRASS. Common throughout in meadows, especially in shade.

POA L.

P. annua L. LOW SPEAR-GRASS. Common in lawns, door-yards and waste places. One of our earliest grasses to mature.

P. compressa L. FLAT-STEMMED POA. WIRE GRASS. ENGLISH BLUE-GRASS. Common in cultivated grounds, along roadsides and in other waste places.

P. flava L. FALSE RED-TOP. FOWL MEADOW GRASS. Plentiful in low meadows and on the banks of streams. (*P. serotina* Ehrh.)

P. pratensis L. KENTUCKY BLUE-GRASS. JUNE GRASS. Common in meadows, fields and woods throughout our range. The commonest of our grasses.

P. trivialis L. ROUGHISH MEADOW-GRASS. Not common, *Fenno; Lucy.*

P. alsodes A. Gray. Rare. Reported by *Lucy, Fenno* and *Coville.*

P. debilis Torr. Frequent in woods, *Fenno.*

PANICULARIA FABR.

P. Canadensis (Michx.) Kuntze. RATTLESNAKE GRASS. TALL QUAKING GRASS. Common in wet places, especially on the borders of swamps. (*Glyceria Canadensis* Trin.)

P. nervata (Willd.) Kuntze. FOWL MEADOW-GRASS. Not uncommon in wet meadows and swamps. (*Glyceria nervata* Trin.)

P. aquatica (L.) Kuntze. REED MEADOW-GRASS. TALL MEADOW-GRASS. Plentiful in marshes or in shallow water. (*Glyceria grandis* Watson.)

P. fluitans (L.) Kuntze. FLOATING MANNA-GRASS. Occasional in shallow water or wet places. (*Glyceria fluitans* R. Br.)

FESTUCA L.

F. ovina L. SHEEP'S FESCUE. Occasional in lawns and waste places.

F. nutans Willd. NODDING FESCUE-GRASS. Frequent in upland woods and rocky places.

F. elatior L. MEADOW FESCUE. Not uncommon in grass-lands throughout our range.

BROMUS L.

B. Kalmii A. Gray. WILD CHESS. Infrequent. Found in hilly or rocky woods, *Lucy; Fenno; Graves.* Elsewhere not reported.

B. secalinus L. CHESS. CHEAT. Tolerably common in fields and waste places.

B. ciliatus L. WOOD CHESS. Frequent in low woodlands and on the banks of streams. The variety *purgans* Gray, by some considered a distinct species, is reported with the type.

LOLIUM L.

L. perenne L. RYE GRASS. Infrequent, *Fenno*. Rare, *Lucy*. Found in dry, waste places.

AGROPYRON J. GÆRTN.

A. repens (L.) Beauv. COUCH GRASS. QUACK GRASS. QUITCH GRASS. Abundant in gardens, fields and waste places throughout our range.

A. caninum (L.) R. & S. AWNED WHEAT GRASS. Not uncommon in dry fields and woods.

HORDEUM L.

H. jubatum L. SQUIRREL-TAIL GRASS. Occasional in waste grounds throughout.

ELYMUS L.

E. Virginicus L. WILD RYE. Tolerably common, especially on the banks of streams.

E. Canadensis L. NODDING WILD RYE. Common on river banks throughout our range.

E. striatus Willd. WILD RYE. Occasional, *Fenno*. Rare, *Clute*. Elsewhere not reported.

HYSTRIX MŒNCH.

H. Hystrix (L.) Millsp. BOTTLE BRUSH GRASS. Occasional. Usually found on the borders of upland woods. (*Asprella Hystrix* Willd.)

EQUISETACEÆ.

EQUISETUM L.

E. arvense L. FIELD HORSETAIL. Very common throughout in any kind of soil. Fertile stems early in spring, brown in color. Sterile, later, much different in appearance from the others.

E. sylvaticum L. WOOD HORSETAIL. Abundant in moist woodlands. Sterile stems producing several very regular whorls of branches. The most beautiful member of the genus in our region.

E. fluviatile L. PIPES. SWAMP HORSETAIL. Plentiful on the borders of lakes, streams and ditches, often in shallow water. Stems frequently unbranched, but usually producing a few straggling branches near the summit of the stem. (*E. limosum* L.)

E. hyemale L. SCOURING RUSH. SHAVE GRASS. Plentiful in low woodlands, on the banks of streams, and other moist places. Stems commonly unbranched, stiff, two feet or more high, persisting through the winter.

E. variegatum Schleich. Rare. Brisbin swamp on the margin of a sand slide, *Coville*. The only station.

FILICES.

POLYPODIUM L.

P. vulgare L. COMMON POLYPODY. Plentiful in our region wherever outcrops of rock occur. Usually found growing on the tops of rocks where there is little moisture. Fronds nearly pinnate, evergreen, leathery; the large, round fruit dots on the under side of the fronds in late summer.

ADIANTUM L.

A. pedatum L. MAIDENHAIR FERN. Abundant throughout in moist, rich woodlands. Stipe a foot or more high, black and shining, divided at the top into two curving branches which bear the pinnæ. Fruit beneath a reflexed portion of the pinnule. Much prized for cultivation.

PTERIS L.

P. aquilina L. BRACKEN. BRAKE. EAGLE FERN. Most abundant on scrubby hillsides, on the borders of fields, roads and in open woods. Stipe stout and tall, at top bearing three divisions, variously subdivided. Rootstock stout, creeping extensively, bearing the fronds at intervals, all summer. Fruit produced in a line on the margin of the pinnules.

PELLÆA Link.

P. gracilis (Michx.) Hook. SLENDER CLIFF-BRAKE. The only station known for this fern within our limits is on a cliff at Killawog, in Broome county. It is there plentiful, *Clute*. Fruit nearly as in *Pteris*.

WOODWARDIA J. E. Smith.

W. Virginica (L.) J. E. Smith. VIRGINIAN CHAIN-FERN. Rare. Brisbin peat-bogs, *Coville*. Thompson's Marsh, *Clute*. Beebe's swamp and swamp at Oakland, *Graves*. Bog north of Barton, *Fenno*. The only stations. Fronds in appearance very much like *Osmunda Cinnamomea*, except that they are borne singly and not in clumps and the fruit dots occur on the backs of fronds that resemble the sterile ones.

ASPLENIUM L.

A. Trichomanes L. MAIDENHAIR SPLEENWORT. WALL SPLEEN-WORT. DWARF SPLEENWORT. Fairly common on the rocky walls of nearly all of our deep ravines. Fronds seldom more than six inches long, pinnate, several from a common centre, forming green rosettes. Fruit dots linear, on the backs of the pinnæ.

A. platyneuron (L.) Oakes. EBONY SPLEENWORT. Common. Found only in rocky uplands in partial or complete shade. Stipe very short, it and the rachis black and polished. Fronds somewhat larger than the preceding, the sterile, half reclining; the fertile, later, erect and longer than the sterile ones. (*A. ebeneum* Ait.)

A. angustifolium Michx. NARROW-LEAVED SPLEENWORT. In rich woods, *Brown*, Van Etten, N. Y., *Barbour*. The only stations. Fronds tall, thin, once pinnate, the fertile ones produced but sparingly.

A. acrostichoides Sw. SILVERY SPLEENWORT. Common in rich, moist shades. Fronds two feet or more high, pinnate, the pinnæ pinnatifid; fruit borne on the back of the fronds, in a double row on each pinnule, silvery white when young. (*A. thelypteroides* Michx.)

A. Filix-fœmina (L.) Bernh. LADY FERN. Abundant along roadsides, in thickets and in open woods. Well known. Quite variable in the form of the fronds. Fruit dots at first curved, at length straight, brown or blackish when ripe.

CAMPTOSORUS Link.

C. rhizophyllus (L.) Link. WALKING FERN. WALKING-LEAF. Rare. Found only in a few restricted areas in our region. Lockwood, N. Y.; scarce, *Barbour*. On rocks west of Barton; rare, *Fenno*. On sandstone rocks in various places about Susquehanna, *Graves*. South Oxford and on sandstone rocks in several places about Oxford, *Coville*. Fronds gradually narrowing from a heart-shaped base, into a long, narrow tip, which bends over and rooting at the apex, gives the common name to the fern. Fruit dots much like *Asplenium*.

PHEGOPTERIS Fee.

P. Phegopteris (L.) Underw. BEECH FERN. Common on wet rocks in sun or shade. Frond triangular in outline, longer than broad, bi-pinnatifid, the lowest pair of pinnæ standing forward. Fruit dots small, round, without indusium. (*P. polypodioides* Fee.)

P. hexagonoptera (Michx.) Fee. SIX-ANGLED POLYPODY. BROAD BEECH FERN. Tolerably common in rather dry, rich woods. In appearance much like the preceding, except that the fronds are broader than long, the lower most pair of pinnæ much larger, their pinnules pinnatifid.

P. Dryopteris (L.) Fee. OAK FERN. Plentiful in rich, moist woods, occasionally in company with *P. Phegopteris*. Fronds small, quite like the bracken in miniature.

DRYOPTERIS Adans.

D. Thelypteris (L.) A. Gray. LADY FERN. MARSH SHIELD FERN. Abundant in swamps, wet woods and other low grounds. Fronds lanceolate, nearly twice pinnate on long stipes. Fruit on the back of fronds late in the season. In shade this fern is usually completely sterile. (*Aspidium Thelypteris* Sw.)

D. Noveboracensis (L.) A. Gray. NEW YORK FERN. Common in moist woods, especially in uplands. Stipe short; frond very thin, tapering both ways from the middle. Distinguished from the preceding, which it much resembles, by the thinner texture of the frond and by the pinnæ gradually narrowing to mere ears at base. (*Aspidium Noveboracensis* Sw.)

D. spinulosa (Retz) Kuntze. SPINULOSE SHIELD FERN. Not common. Found throughout, but much less common than the following sub-species. Fronds tall, ovate, bi pinnate, the pinnules with spinulose teeth. The fronds are produced in early spring, several in a clump and are half evergreen. (*Aspidium spinulosum* Sw.)

D. spinulosa intermedia (Muhl.) Underw. SPINULOSE SHIELD-FERN. One of the commonest of ferns. Found in woodlands wet or dry. Resembles *D. spinulosa*, but with the pinnæ and pinnules more crowded and finely dissected. (*A. spinulosum*, var. *intermedium* D. C. Eaton.)

D. spinulosa dilatata (Hoffm.) Underw. Much less common than the type and usually found at higher elevations. (*Aspidium spinulosum*, var. *dilatatum* Hook.)

——— It is a difficult matter for the beginner to properly separate *Dryopteris spinulosa* from its varieties. The fronds vary somewhat in shape, and it is only with an abundance of material that one can distinguish the limits of the forms.

D. Boottii (Tuckerm.) Underw. BOOTT'S SHIELD FERN. Rare. Susquehanna, *Graves*. Pond Brook. *Clute*. The only stations reported. In appearance this fern is about half way between *D. spinulosa* and the following. Further study of the forms referred to these two ferns will doubtless discover more stations for the present species. (*Aspidium Boottii* Tuckerm.)

D. cristata (L.) A. Gray. CREST FERN. CRESTED SHIELD-FERN. Plentiful in swampy woods. Fronds tall and narrow, lanceolate in outline, pinnate, their divisions pinnatifid, the lowest pair of pinnæ triangular. The sterile fronds are usually shorter and broader than the fertile and less erect. (*Aspidium cristatum* Sw.)

D. cristata Clintoniana (D. C. Eaton) Underw. Rare. Swamp near the city of Binghamton, *Clute* Unadilla Forks, *Brown*. The only stations recorded. The fronds of this fern are in every way larger than the preceding, which it otherwise much resembles. Closer observation of our ferns will probably bring to light other stations for this one. (*Aspidium cristatum*, var. *Clintonianum* D. C. Eaton.)

D. Goldieana (Hook) A. Gray. GOLDIE'S SHIELD-FERN. Rare. In swampy ground east of Van Etten. Chemung county, N. Y., *Barbour*. Near Unadilla Forks; not common, *Brown*. The only stations. This fern may occur at other points within our limits and should be looked for in cold, wet woods. In appearance it is considerably like the following and might be mistaken for a large form of it. (*Aspidium Goldieana* Hook.)

D. marginalis (L.) A. Gray. MARGINAL SHIELD-FERN. EVERGREEN WOOD-FERN. Abundant in deep shades. Fronds two feet or more high, about twice pinnate. Well known for the bright, white covering to the fruit dots, which are borne on the back of fronds like the sterile ones. The fronds remain green through the winter, but are usually prostrated by wind and snow. (*Aspidium marginale* Swartz.)

D. acrostichoides (Michx.) Kuntze. CHRISTMAS FERN. Abundant throughout our range, especially in evergreen woods. Fronds thick, evergreen, pinnate, the fertile contracted at the summit. Among the earliest of our ferns to fruit. The variety *incisa* is occasionally found with the type, and differs from it in having the pinnæ much incised and in fertile fronds, nearly all of them fruit-bearing. This species is one of our best known ferns and is much used for holiday decorations. (*Aspidium acrostichoides* Sw.)

CYSTOPTERIS Bernh.

C. bulbifera (L.) Bernh. BULB-BEARING BLADDER-FERN. Somewhat rare. Oxford, *Coville*. Killawog, N. Y., *Clute*. Cascade, near Susquehanna, *Graves*. North of Apalachin, *Fenno*. Ravines at Unadilla Forks, *Brown*. Elsewhere not reported. Found usually on cliffs. Fronds very long and narrow, prostrate or pendant, bi-pinnate. In addition to fruit dots, the fronds usually bear bulblets on the backs.

C. fragilis (L.) Bernh. FRAGILE BLADDER-FERN. BRITTLE FERN. Common in wet, shaded soil, especially on cliffs. Fronds rather slender, nearly a foot long, twice or thrice pinnate.

ONOCLEA L.

O. sensibilis L. SENSITIVE FERN. Very abundant. Found on the borders of ponds and streams and in all low grounds. Sterile fronds broad and coarse; fertile fronds later in the season, their segments contracted into berry-like bodies which enclose the fruit. The fertile fronds remain erect through the winter. The so-called variety *obtusilobata* is occasionally found. In appearance it is half-way between the fertile and sterile fronds of this species.

O. Struthiopteris (L.) Hoffm. OSTRICH FERN. Abundant. Found in alluvial soil along our larger streams. Fronds in a circular clump, often twenty or more from the same root, pinnate, the divisions pinnatifid. The fertile fronds appear late in summer from the midst of the sterile ones, much different in appearance. The rootstock produces long, slender stolons, which form new plants at their ends. Our tallest and handsomest fern.

WOODSIA R. Br.

W. Ilvensis (L.) R. Br. RUSTY POLYPOD. Exposed rocks, summit of Mt. Markham, Unadilla Forks, *Brown*. The only station.

W. obtusa (Spreng.) Torr. OBTUSE WOODSIA. Rare. Found occasionally on the rocky walls along the Susquehanna below Towanda, *Clute*. Elsewhere not reported. In aspect this is much like *Cystopteris fragilis*, and at first glance might be mistaken for that species.

DICKSONIA L'Her.

D. punctilobula (Michx.) A. Gray HAY-SCENTED FERN. FINE-HAIRED MOUNTAIN FERN. Plentiful. Found in open woods and thickets especially in uplands. Fronds two or three feet long, very finely cut, usually growing in dense patches. Fragrant in drying. (*D. pilosiuscula* Willd.)

LYGODIUM Sw.

L. palmatum (Bernh.) Sw. CLIMBING FERN. "McDonough, Chenango county (*Mrs. D. B. Fitch*.) This is the second station in which this fern has been found in our State."—*Annual Report State Botanist, 1893.*

OSMUNDA L.

O. regalis L. FLOWERING FERN. ROYAL FERN. Plentiful in swamps and wet, open woodlands, often growing in shallow water. Fronds very smooth, twice pinnate, the pinnules oblong. Fruit borne in a panicle at the summit of some of the fronds.

An elegant species, much less like the popular conception of a fern than most of our common species.

O. Claytoniana L. INTERRUPTED FLOWERING FERN. CLAYTON'S FERN. Common in low grounds, especially in thickets along streams. Sterile fronds once pinnate, the pinnæ pinnatifid. Fruit borne on two or more pairs of much contracted pinnæ in the middle of an otherwise unchanged frond. This species fruits as the fronds unroll.

O. cinnamomea L. CINNAMON FERN. Very common in all low grounds. Fronds in a circular clump, the sterile pinnate with the pinnæ pinnatifid, the fertile fronds much contracted, covered with cinnamon-colored sporangia, appearing earlier than the sterile ones.

OPHIOGLOSSACEÆ.

BOTRYCHIUM Sw.

B. simplex E. Hitchcock. MOONWORT. Swamp at the head of Bennett's Lane, Oxford, 1886, *Coville*. Unadilla Forks; rare, *Brown*. The only stations.

B. lanceolatum (S. G. Gmel.) Angs. LANCEOLATE MOONWORT. Woods at base of Mt. Markham, *Brown*. In same situations as the following and almost invariably accompanying it, *Coville*.

B. matricariæfolium A. Br. Rare. Woods at base of Mt. Markham, *Brown*. Beech and maple woods at various points in Oxford and Preston, *Coville*.

B. ternatum (Thunb.) Sw. TERNATE MOONWORT. GRAPE FERN. Plentiful on dryish knolls. Sterile portion of the plant broadly triangular, ternate and variously divided. Fertile portion erect, often four times pinnate. Our common form of the plant may be referred to the variety *obliquum*. A form with the segments cut into narrow lobes or teeth is occasionally found and is the variety *dissectum*.

B. Virginianum (L.) Sw. RATTLESNAKE FERN. GRAPE FERN. VIRGINIAN MOONWORT. Common in rich, moist woods throughout our range. In appearance like the preceding, except that the plant is taller with thinner fronds. It is found much earlier in the season than *B. ternatum*.

OPHIOGLOSSUM L.

O. vulgatum L. ADDER'S-TONGUE FERN. Occasional in moist, grassy openings of woods in northern Chenango county, *Coville*.

LYCOPODIACEÆ.

LYCOPODIUM L.

L. lucidulum Michx. SHINING CLUB-MOSS. HEMLOCK CLUB-MOSS. Common in cold dark, woodlands. In appearance likened to the hemlock

L. annotinum L. STIFF CLUB-MOSS. Rare. Balsam swamp, Pharsalia, *Coville*. Along Pierce Creek, two miles south of Binghamton, *Clute*. Low ground, one mile north of Apalachin, *Fenno*. The only recorded stations.

L. obscurum L. GROUND PINE. Plentiful in moist, rich woods. In appearance somewhat like a miniature pine tree, especially the variety *dendroideum*, which frequently occurs.

L. clavatum L. COMMON CLUB-MOSS. RUNNING PINE. Plentiful throughout. Found in open woods, thickets, and along bushy roadsides. Stem creeping extensively, with similar ascending leafy branches; fruit borne in conspicuous cylindrical spikes, two or three on a slender peduncle.

L. complanatum L. GROUND PINE. RUNNING PINE. CEDAR CLUB-MOSS. Common in rather dry soil. In appearance this species has been likened to the cedar. Fruit as in the preceding. The variety *chamæcyparissus*, with narrower and more erect branches, is occasionally found with the type.

SELAGINELLACEÆ.

SELAGINELLA BEAUV.

S. apus (L.) Spring. CREEPING SELAGINELLA. Common in pastures by the river at Unadilla Forks, *Brown*. The only station. This species very much resembles a moss, and has doubtless been overlooked in other parts of our range.

ISOETACEÆ.

ISOETES L.

1. **Engelmanni** A. Br. Along the river at Apalachin, *Fenno.* Along the Susquehanna in gravelly soil, city of Binghamton, *Clute.* Elsewhere not reported. Plant sedge-like in appearance, and difficult to distinguish from the vegetation amid which it grows.

1. **Engelmanni gracilis** Engelm. Found with the type in very wet, shady places. Leaves few and very slender.

INDEX TO THE GENERA.

Synonyms in Italic.

	PAGE		PAGE
Abies	102	Aster	56
Abutilon	20	Astragalus	30
Acalypha	93	Atragene	1
Acer	25	Atriplex	84
Achillea	61	Avena	130
Aconitum	5	Azalea	64
Acorus	116		
Acroanthes	102	Baptisia	28
Actæa	5	Barbarea	12
Adiantum	134	Batrachium	3
Adicea	94	Benzoin	92
Adlumia	10	Berberis	7
Æsculus	25	Betula	7
Agrimonia	37	Bicuculla	10
Agropyron	133	Bidens	61
Agrostemma	17	Blephilia	86
Agrostis	130	Bœhmeria	94
Ailanthus	22	Botrychium	140
Alisma	117	Brachyelytrum	129
Allium	108	Brasenia	7
Alnus	97	Brassica	13
Alopecurus	129	Bromus	132
Alsine	18	*Brunella*	87
Amaranthus	88	Bursa	13
Ambrosia	59		
Amelanchier	38	Cacalia	62
Ampelopsis	25	Calamagrostis	130
Amphicarpæa	32	*Calamintha*	86
Anagallis	73	Calla	115
Anaphalis	58	Callitriche	42
Andromeda	68	*Calopogon*	104
Andropogon	128	Caltha	4
Anemone	1	Campanula	66
Anemonella	2	Camptosorus	136
Angelica	45	Cannabis	94
Antennaria	58	Capnoides	10
Anthemis	61	*Capsella*	13
Anthoxanthum	128	Cardamine	11
Anychia	88	Carduus	3
Aphyllon	83	Carex	121
Apios	31	Carpinus	97
Apocynum	74	Carum	46
Aquilegia	5	*Carya*	5
Arabis	11	*Cassandra*	6
Aralia	48	Cassia	32
Arceuthobium	92	Castalia	8
Arctium	62	Castanea	96
Arenaria	17	Castilleja	82
Arisæma	115	Catalpa	4
Aronia	38	Caulophyllum	7
Artemisia	62	Ceanothus	24
Asarum	91	Celastrus	24
Asclepias	74	Celtis	94
Asparagus	109	Cephalanthus	51
Aspidium	137	Cerastium	18
Asplenium	135	Ceratophyllum	101
Asprella	133	Chærophyllum	47

	PAGE		PAGE
Chamædaphne	69	Disporum	110
Chamælirium	112	Drosera	41
Chamænerion	43	Dryopteris	136
Chamæraphis	128	Dulichium	119
Chelidonium	9		
Chelone	80	Eatonia	131
Chenopodium	89	*Echinocystis*	45
Chimaphila	70	*Echinospermum*	76
Chiogenes	68	Echium	77
Chrysanthemum	61	Eleocharus	119
Chrysopogon	128	*Elodea*	102
Chrysosplenium	39	*Elodes*	19
Cichorium	63	Elymus	133
Cicuta	47	Epigæa	68
Cimicifuga	5	Epilobium	43
Cinna	130	Epiphegus	83
Circæa	44	Equisetum	133
Cladium	121	Eragrostis	131
Claytonia	18	Erechtites	62
Clematis	1	Erigenia	47
Clinopodium	86	Erigeron	58
Clintonia	110	Eriocaulon	118
Cnicus	63	Eriophorum	120
Collinsonia	85	Erysimum	13
Comandra	92	Erythronium	111
Comarum	36	Euonymus	23
Comptonia	96	Eupatorium	54
Conium	47	Euphorbia	92
Conopholis	83	Euthamia	56
Convallaria	109		
Convolvulus	77	Fagopyrum	91
Coptis	5	Fagus	99
Corallorhiza	103	Falcata	31
Cornus	49	Festuca	132
Coronilla	30	Fragaria	35
Corydalis	10	Fraxinus	73
Corylus	97	Fumaria	10
Cratægus	38		
Cryptotænia	46	Galeopsis	87
Cunila	85	Galinsoga	61
Cuscuta	78	Galium	53
Cynoglossum	76	Gaultheria	68
Cypripedium	106	Gaura	44
Cyperus	110	Gaylussacia	66
Cystopteris	138	Gentiana	75
		Geranium	21
Dactylis	131	Gerardia	82
Dalibarda	34	Geum	35
Danthonia	131	*Gillenia*	33
Daphne	92	Glecoma	87
Dasystoma	82	Gleditschia	32
Datura	79	*Glyceria*	132
Daucus	45	Gnaphalium	59
Decodon	43	*Goodyera*	104
Dentaria	10	Gratiola	81
Deringa	46	Gyrostachys	104
Deschampsia	130		
Desmodium	30	Habenaria	105
Dianthera	84	Hamamelis	42
Dianthus	16	Hedeoma	86
Dicentra	10	Helenium	61
Dicksonia	139	Helianthemum	14
Diervilla	52	Helianthus	60
Dioscorea	108	Heliopsis	60
Dipsacus	54	Hemerocallis	109
Dirca	92	Hepatica	2

	PAGE
Heracleum	45
Hesperis	12
Heteranthera	114
Hibiscus	20
Hicoria	95
Hieracium	64
Holcus	130
Homalocenchrus	128
Hordeum	131
Houstonia	52
Humulus	94
Hydrangea	40
Hydrocotyle	48
Hydrophyllum	70
Hypericum	19
Hypopitys	72
Hypoxis	107
Hystrix	133
Ilex	23
Ilicioides	23
Ilysanthes	81
Impatiens	22
Inula	59
Ipomœa	78
Iris	107
Isoetes	142
Jeffersonia	7
Juglans	95
Juncoides	114
Juncus	113
Juniperus	102
Kalmia	69
Kneiffia	44
Kœleria	131
Kœllia	85
Lactuca	65
Lamium	87
Lappula	76
Laportea	94
Larix	102
Lathyrus	31
Lechea	14
Ledum	70
Leersia	128
Legouzia	66
Lemna	116
Leonurus	87
Lepidium	13
Leptandra	81
Leptorchis	103
Lespedeza	31
Ligustrum	73
Lilium	111
Limodorum	104
Linaria	80
Lindera	92
Linnæa	51
Linum	20
Liparis	103
Liriodendron	6
Listera	104
Lithospermum	77
Lobelia	65

	PAGE
Lolium	113
Lonicera	52
Lophanthus	87
Ludwigia	43
Lupinus	28
Luzula	114
Lychnis	17
Lycium	79
Lycopodium	141
Lycopsis	77
Lycopus	85
Lygodium	139
Lysimachia	72
Lysimachia	73
Magnolia	6
Maianthemum	110
Malva	20
Medeola	112
Medicago	27
Meibomia	30
Melampyrum	83
Melilotus	29
Menispermum	6
Mentha	85
Mertensia	76
Menyanthes	75
Micrampelis	44
Microstylis	103
Milium	129
Mimulus	80
Mitchella	53
Mitella	40
Mollugo	45
Monarda	86
Moneses	71
Monotropa	71
Monotropa	72
Morus	94
Muhlenbergia	127
Myosotis	77
Myrica	96
Myriophyllum	42
Naias	118
Nasturtium	12
Naumburgia	73
Negundo	26
Nemopanthes	23
Nepeta	87
Nymphæa	8
Nyssa	50
Nuphar	8
Oakesia	111
Œnothera	43, 44
Onagra	43
Onoclea	138
Onosmodium	77
Ophioglossum	141
Opulaster	33
Orchis	105
Origanum	86
Orontium	116
Oryzopsis	129
Osmorrhiza	47
Osmunda	139

	PAGE		PAGE
Ostrya	97	Rhodora	70
Oxalis	21	Rhus	26
		Rhynchospora	121
Panax	49	Ribes	40
Panicularia	132	Robinia	29
Panicum	127	Roripa	12
Papaver	9	Rosa	37
Parnassia	40	Rotala	42
Parthenocissus	25	Rubus	34
Pastinaca	45	Rudbeckia	60
Pedicularis	82	Rumex	89
Pellæa	135		
Peltandra	115	Sagittaria	117
Penthorum	41	Salix	99
Pentstemon	80	Sambucus	50
Peramium	104	Sanguinaria	9
Phalaris	128	Sanguisorba	37
Phegopteris	136	Sanicula	48
Phleum	129	Saponaria	16
Phlox	75	Sarracenia	9
Phryma	84	Sassafras	91
Physalis	79	Saxifraga	39
Physalodes	79	Scabiosa	54
Physocarpus	33	Scheuchzeria	117
Phytolacca	89	Schollera	67
Picea	101	Scirpus	120
Pilea	94	Scrophularia	80
Pimpinella	46	Scutellaria	87
Pinguicula	83	Sedum	41
Pinus	101	Selaginella	141
Plantago	88	Senecio	62
Platanus	95	Sericocarpus	56
Poa	131	*Setaria*	128
Podophyllum	7	Sicyos	44
Pogonia	104	Silene	16
Polanisia	14	Sinapis	13
Polemonium	76	Sisymbrium	13
Polygala	27	Sisyrinchium	107
Polygonatum	109	Sium	46
Polygonum	90	*Smilacina*	110
Polymnia	59	Smilax	108
Polypodium	134	Solanum	78
Pontederia	113	Solidago	54
Populus	100	Sorbus	38
Porteranthus	33	Sparganium	114
Poterium	37	Spathyema	115
Portulacca	18	*Specularia*	66
Potamogeton	117	Spergula	18
Potentilla	36	Spiræa	33
Prenanthes	64	*Spiranthes*	104
Prunella	87	Spirodela	116
Prunus	32	Sporobolus	130
Ptelea	22	Stachys	88
Pteris	134	Staphylea	26
Pterospora	71	Steironema	72
Pycnanthemum	85	*Stellaria*	18
Pyrola	71	Streptopus	110
Pyrus	37	Symphoricarpos	52
		Symphytum	77
Quercus	98	*Symplocarpus*	115
		Syndesmon	2
Ranunculus	3		
Raphnus	13	Tanacetum	62
Razoumofskya	92	Taraxicum	64
Rhamnus	24	Taxus	102
Rhododendron	70	Teucrium	85

	PAGE		PAGE
Thalictrum	2	Vagnera	109
Thaspium	46	Valerianella	53
Thuja	102	Vallisneria	103
Tiarella	39	Veratrum	113
Tilia	20	Verbascum	79
Tradescantia	113	Verbena	84
Tragopogon	63	Veronica	81
Trichostema	84	Viburnum	50
Trientalis	72	Vicia	31
Trifolium	28	Vinca	73
Trillium	112	Viola	14
Triosteum	51	Vitis	24
Tsuga	101	Vleckia	86
Tussilago	62		
Typha	114	Waldsteinia	35
		Woodsia	137
Udora	102	Woodwardia	135
Ulmus	93		
Unifolium	110	Xanthium	57
Urtica	94	Xanthorrhiza	6
Urticastrum	94	Xanthoxylum	22
Utricularia	83	Xolisma	68
Uvularia	110		
		Zannichellia	118
Vaccinium	67	Zizia	46

INDEX TO COMMON NAMES.

	PAGE		PAGE
Abele	100	Barberry	7
Acacia	29, 32	Barnyard-grass	127
Aconite	5	Basswood	20
Adder's-mouth	103	Beak-rush	121
Adder's-tongue	111	Bean	31, 34
Adder's-tongue Fern	141	Beard-grass	128
Agrimony	37	Beard-tongue	80
Alder	23, 97	Bear's Ears	106
Alexanders	46	Beaver-poison	47
Alfalfa	29	Bedstraw	53
Alleghany-vine	10	Bee-balm	86
Allspice	92	Beech	97, 97
Amaranth	88, 89	Beechdrops	72, 83
Anemone	1, 2	Beech Fern	136
Angelica-tree	48	Beggar-lice	61, 76
Apple	38	Beggar-ticks	61
Arbor-vitæ	102	Bellflower	66
Arbutus	68	Bellwort	110, 111
Arrow Arum	115	Benjamin-bush	92
Arrow-head	117	Bent grass	130
Arrow-wood	51	Bergamot	85, 86
Artichoke	60	Betony	92
Arum	115	Bindweed	77, 78, 91
Ash	22, 38, 73	Birch	68, 96
Aspen	100	Bird's-nest	45, 72
Aster	56, 57, 58	Birth-root	112
Avens	35	Bishop's-cap	37
Azalea	69	Bitternut	96
Baby-foot	27	Bittersweet	23, 78
Balm	85, 86	Bitter-weed	57
Balm-of-Gilead	101	Blackberry	34
Balsam-apple	44	Black-bur	55
Baneberry	5, 6	Black-cap	34

	PAGE		PAGE
Bladder Fern	138	Cat-tail	114
Bladder Ketmia	20	Cedar	102
Bladder-nut	26	Celandine	9
Bladder-wort	83	Chain Fern	135
Blazing Star	112	Chamomile	61
Blite	89	Charlock	13
Bloodroot	9	Cheat	133
Bluebell	66	Checkerberry	68
Blueberry	67	Cherry	32, 33
Blue-curls	87	Chervil	47
Bluets	52	Cheeses	20
Blue-eyed grass	107	Chess	133
Blue-grass	132	Chestnut	99
Blue-joint	130	Chickweed	18, 45, 88
Blue-stem	128	Chickory	63
Blue-weed	77	Chokeberry	38
Boneset	54	Christmas Fern	138
Bottle brush-grass	133	Cinnamon Fern	140
Bouncing-Bet	16	Cinquefoil	36
Bowman's-root	33	Clayton's Fern	140
Box Elder	26	Clearweed	94
Bracken	135	Cleavers	53
Brake	135	Clematis	1
Brittle Fern	138	Cliff-brake	135
Brooklime	81	Climbing Fern	139
Brown-eyed Susan	60	Clot-bur	59
Buckbean	75	Clover	28, 29, 31
Buckthorn	24	Club Moss	141
Buckwheat	91	Club-rush	120
Bugbane	5	Cockle-bur	59
Bugle-weed	85	Cockspur-grass	127
Bugloss	77	Coffee	51
Bullrush	113, 120	Cohosh	5, 6, 7
Bunch-berry	49	Coltsfoot	62, 91
Bur-cucumber	44	Columbine	5
Burdock	62	Comfrey	76, 77
Bur-flower	76	Cone-flower	60
Bur Marigold	61	Coral-root	103
Burnet	37	Corn-cockle	17
Bur-reed	114, 115	Cornel	49, 50
Bur-seed	76	Corn Spurry	18
Bush Clover	31	Corpse-plant	71
Butter-and-eggs	80	Cotton-grass	120
Buttercup	3, 4	Cowbane	47
Butterfly-weed	74	Cow-lily	8
Butternut	95	Cowslip	4, 76
Butter-weed	58, 62	Cow-wheat	83
Butter-wort	83	Couch-grass	133
Bu'tonball	95	Crab-apple	37
Button-bush	52	Crab-grass	127
Buttonwood	95	Cranberry	51, 67
		Cranberry Tree	51
Calamint	86	Cranesbill	21
Calamus-root	116	Crawley-root	103
Calico-bush	69	Cress	11, 12, 13
Campion	16, 17	Crest Fern	137
Canary-grass	128	Crinkle-root	10
Cancer-root	83	Crowfoot	3, 4
Caraway	46	Cuckoo-flower	11
Cardinal-flower	65	Cucumber	44
Carpet-weed	45	Cucumber-root	11
Carrion-flower	108	Cucumber-tree	6
Carrot	45	Cudweed	57
Cat-briar	108	Culver's Physic	81
Catchfly	17	Currant	40
Catnip	87	Cut-grass	128

INDEX.

	PAGE		PAGE
Daisy	53, 60, 61	Go-to-bed-at-noon	63
Dame's Violet	12	Grape	24
Dandelion	64	Grape Fern	140
Dead-nettle	87	Grass-of-Parnassus	40
Deerberry	67	Graveyard-weed	43
Dewberry	34	Green-briar	138
Devil's-bit	112	Green-dragon	115
Devil's-needles	61	Gromwell	77
Ditch-moss	102	Ground-cherry	74
Dittany	85	Ground-ivy	87
Dock	89, 90	Ground-nut	31, 49
Dockmackie	51	Ground-pine	141
Dodder	78	Groundsel	62
Dog-bane	74	Gum	50
Dog-fennel	61		
Dog-tooth Violet	111	Hackberry	94
Dogwood	25, 49	Hackmetack	102
Dropflower	64	Hair-grass	130
Drop-seed	129	Harbinger-of-spring	47
Duckmeat	116	Hard-hack	33
Dutchman's Breeches	10	Hare-bell	66
		Haw	38, 51
Eagle Fern	135	Hawkweed	64
Eel-grass	103	Hay-scented Fern	139
Eglantine	37	Hazel	42
Elder	49, 50	Hazelnut	97
Elecampane	59	Heartsease	16
Elm	93, 94	Heart-weed	90
Everlasting	58	Hellebore	113
Eye-berry	53	Hemlock	47, 101, 102
		Hemp	74
False Violet	34	Hemp-nettle	87
Fescue	132	Henbit	87
Fever-bush	92	Hepatica	2
Fever-wort	39, 51	Herb-Robert	21
Fig-wort	80	Hercules' Club	48
Finger-grass	127	Herds-grass	129
Fir	102	Hickory	95, 96
Fire-weed	43, 62	Hobble-bush	50
Five-finger	36	Hogweed	59
Flag	107, 116	Holly	23
Flax	20	Honeysuckle	5, 52, 84
Fleabane	58	Honewort	46
Flowering Fern	140	Hop	94
Forget-me-not	52, 77	Hop-tree	22
Fox-tail	128, 129	Horehound	85
Foxglove	82	Hornbeam	97
Frog-lily	8	Hornwort	101
Frost-weed	14	Horse-chestnut	25
Fumitory	10	Horse-radish	12
		Horse-tail	134
Gall-of-the-earth	64, 106	Horse-weed	58
Garden Orpine	41	Hound's-tongue	76
Garget	89	Huckleberry	38, 66, 67
Garlic	109	Huntsman's-cup	9
Gentian	51, 75	Hyssop	81, 86
Germander	85		
Gill-over-the-Ground	87	Indian Bean	84
Ginger	91	Indian-grass	128
Ginseng	49	Indian Physic	33
Goatsbeard	63	Indian Pipe	71
Golden Club	116	Indian Poke	114
Goldenrod	54, 55, 56	Indian Turnip	115
Goldthread	5	Indigo	28
Gooseberry	40	Inkberry	80
Goose grass	36, 53	Innocents	52
Goosefoot	89		

	PAGE
Ironwood	97
Ivy	27
Jacob's Ladder	76, 80
Jack-in-the-Pulpit	115
Jerusalem-oak	89
Jewel-weed	22
Jimson-weed	79
Joe-pye-weed	54
Juneberry	38
June-grass	132
Kingnut	95
Kinnikinnik	49
Knot-grass	90
Lady's-slipper	106, 107
Lady's-tresses	104
Lady Fern	136
Lady's-smock	11
Lady's-thumb	90
Lamb-kill	69
Lamb's-lettuce	53, 54
Lamb's-quarters	89
Larch	102
Laurel	68, 69, 70
Leaf-cup	59
Leather-leaf	69
Leather-wood	92
Leek	108
Lettuce	64, 65
Lever-wood	97
Lily	109, 111, 112
Linden	20
Lion's-foot	64
Liquorice	53
Live-for-ever	41
Liverwort	2
Locust	29, 32
Loose-strife	43, 72, 73
Lop-seed	84
Lousewort	82
Love-vine	78
Lucerne	29
Lungwort	77
Lupine	28
Maidenhair Fern	134
Mallow	20
Maltese Cross	17
Mandrake	7
Manna-grass	132
Maple	25, 26
Maplebush	25
Mare's-tail	58
Marigold	61
Marjoram	86
Marsh-marigold	4
Matrimony-vine	79
May-apple	7
May-flower	68, 69
Mayweed	61
Meadow-grass	130
Meadow-rue	2
Meadow-sweet	33
Medic	29
Melilot	29

	PAGE
Mercury	93
Milfoil	61
Milkweed	65, 74
Milk-wort	28
Millet	128, 129
Mint	85
Mitre-wort	39
Moccasin-flower	107
Mockernut	95
Moneywort	73
Monkey-flower	80
Monks-hood	5
Moonseed	6
Moonwort	140
Moosewood	25, 92
Morning-glory	78
Motherwort	87
Mountain Fringe	10
Mountain Fern	139
Mouth-root	5
Mulberry	34
Mullein	79, 80
Mustard	12, 13
Musquash-root	47
Myrtle	73
Nannyberry	51
Neck-weed	81
Nettle	88, 94
Nettle Tree	94
New Jersey Tea	24
New York Fern	137
Nightcaps	1
Nightshade	44, 78, 79
Nimble-will	129
Nine-bark	33
None-such	29
Nut-grass	119
Oak	98
Oak Fern	136
Oats	111, 130
Oat-grass	131
Old Witch-grass	127
Onion	108, 109
Orchard-grass	131
Orchis	103, 105, 106
Osier	49
Ostrich Fern	139
Oswego Tea	86
Oxeye	60, 61
Painted Cup	82
Pansy	16
Parsnip	45, 46
Partridge-berry	53
Pea	31
Peanut	31
Pennyroyal	84, 86
Penny-wort	13, 48
Pepper-and-salt	47
Pepper-grass	13
Pepperidge	50
Peppermint	85
Pepper-root	10, 11
Periwinkle	73
Pettymorrel	48

	PAGE
Pickerel-weed	113
Pigeon-berry	83
Pigeon-grass	128
Pignut	76
Pig-weed	89
Pimpernel	7, 81
Pine	101
Pine-drops	71
Pine-sap	71, 72
Pink	17, 27, 52, 69, 75, 104
Pinkster	69
Pin-weed	14
Pipes	134
Pipewort	118
Pipsissiwa	70
Pitcher-plant	9
Pitch-forks	61
Piunkum	62
Plane-tree	95
Plantain	58, 62, 88, 104, 113, 117
Pleurisy-root	74
Plum	32
Poison Oak	27
Pokeweed	89
Polypody	134, 136, 139
Pond-lily	8
Pondweed	117, 118
Poplar	6, 100, 101
Prim	73
Primrose	43, 44
Princes-feather	90
Princess-pine	70
Privet	73
Purslane	18, 43, 92
Quack-grass	133
Quaking-grass	132
Queen Anne's Lace	45
Quercitron	98
Quitch-grass	133
Raccoon-berry	7
Ragweed	59
Ragwort	62
Ramstead	80
Raspberry	34
Rattlesnake Fern	140
Rattlesnake-grass	132
Rattlesnake Plantain	104
Rattlesnake Root	64
Rattlesnake Weed	64
Red-root	24, 89
Redtop	130, 132
Rheumatism-root	7, 80
Rhododendron	70
Rice	128, 129
Rich-weed	85, 94
Robin's Plantain	58
Rocket	12
Rock Rose	14
Rose	37
Rosemary	68
Royal Fern	140
Running-pine	141
Rush	113, 114, 119, 134
Rutland Beauty	78

	PAGE
Rye-grass	131
Salsify	63
Sand-wort	17
Sanicle	48
Sarsaparilla	48
Sassafras	91
Savin	101
Saxifrage	33
Scabious	58
Scouring-rush	134
Scurvy-grass	12
Sedge	121-126
Self-heal	87
Senna	32
Sensitive Fern	139
Serviceberry	38
Shad-bush	38
Shagbark	95
Shamrock	29
Shave-grass	134
Sheepberry	51
Shellbark	95
Shepherd's Purse	13
Shield Fern	136, 137, 139
Shinleaf	71
Short-husk	132
Sickle-pod	11
Side-saddle Flower	9
Silkweed	74
Silver-rod	55
Silverweed	32, 36
Skull-cap	81, 87
Skunk-cabbage	115
Sloe	51
Smartweed	90
Snake-head	83
Snake-root	5, 28, 48, 54, 91
Snakeweed	102
Snapdragon	22
Snapweed	22
Sneezeweed	61
Snowberry	52, 63
Soapwort	16
Solomon's-seal	109, 110
Sorrel	21, 90
Sour-grass	21
Sour Gum	50
Sow-thistle	63
Spanish-needles	61
Spatterdock	8
Spear-grass	132
Spearwort	3
Speedwell	81
Spicewood	92
Spiderwort	111
Spignet	48
Spikenard	48, 124
Spike-rush	119, 120
Spleenwort	135
Spoon-wood	69
Spring Beauty	18, 17
Spearmint	85
Spruce	101
Spurge	42, 93
Squawberry	53

INDEX.

	PAGE
Squaw-root	83
Squaw-weed	62
Squirrel-cups	2
Squirrel-corn	10
Squirrel-grass	133
Staff-tree	23
Star-flower	72
Star-grass	107, 113
Starwort	42
Steeple-bush	33
Stick-tight	61
Stickweed	76
Stitchwort	18
St. John's-wort	19
Stonecrop	41
Stoneroot	85
Strawberry	35
Strawberry-bush	23
Succory	63
Sugarberry	94
Sumac	26, 27
Sundew	41
Sundial	28
Sunflower	60
Sundrops	44
Swallow-wort	9
Sweet-briar	37
Sweet Cicely	47
Sweet Clover	29
Sweet Fern	96
Sweet Flag	116
Sweet William	16
Sycamore	95
Tamarack	102
Tansy	62
Tape-grass	102
Tare	31
Teaberry	53, 68
Tear-thumb	91
Teasel	54
Thimble-berry	34
Thimble-weed	1
Thin-grass	130
Thistle	63
Thorn	38
Thornapple	79
Thoroughwort	54
Three-leaved Mercury	27
Timothy	129
Tinker's-weed	51
Toadflax	80, 92
Tobacco	66
Tomato	79
Toothache-tree	22
Tooth-wort	10
Touch-me-not	22
Traveller's Joy	1
Tree-moss	93
Trefoil	22, 30
Trillium	112
Trumpet-weed	54
Tulip-tree	6
Tumble-weed	89
Tupelo	50
Turnip	115

	PAGE
Turtle-head	80
Tway-blade	103, 104
Twig-rush	121
Twin-flower	51
Twin-leaf	7
Twisted-stalk	110
Valerian	76
Vegetable Oyster	63
Velvet-grass	130
Velvet Leaf	20
Venus' Looking-glass	66
Vernal-grass	128
Vervain	84
Vetch	30, 31
Vetchling	31
Violet	12, 14, 15, 16, 34, 111
Virgin's-bower	1
Virginia-creeper	25, 78
Wake-robin	112
Walking-fern	136
Walking-leaf	136
Walnut	95
Wartweed	93
Water-beech	97
Water-carpet	39
Water-leaf	76
Water-lily	8
Water-nymph	8
Water-pepper	90
Water-shield	7
Water-target	7
Water-willow	43, 84
Waxwork	23
Wayfaring Tree	50
Wheat-grass	133
Wheat-thief	77
Whistlewood	25
White-root	48
White-weed	58, 61
Whitewood	6, 20
Wild Lemon	7
Wild Oats	111
Willow	43, 84, 99, 100
Willow-herb	43
Wind-flower	1
Winterberry	23
Wintergreen	27, 68, 71, 72
Wire-grass	131, 132
Witch-hazel	42
Witch-hopple	50
Withe Rod	51
Wolfbane	5
Woodbine	25
Wood Fern	138
Wood-grass	128
Wood-sage	85
Wood-sorrel	21
Wool-grass	120
Wormwood	59
Yam	108
Yarrow	61
Yellow-root	6
Yellow-seed	13
Yew	102

www.ingramcontent.com/pod-product-compliance
Lightning Source LLC
Chambersburg PA
CBHW020301170426
43202CB00008B/457